Lecture Notes in Electrical Engineering

Volume 412

About this Series

"Lecture Notes in Electrical Engineering (LNEE)" is a book series which reports the latest research and developments in Electrical Engineering, namely:

- Communication, Networks, and Information Theory
- Computer Engineering
- Signal, Image, Speech and Information Processing
- Circuits and Systems
- Bioengineering

LNEE publishes authored monographs and contributed volumes which present cutting edge research information as well as new perspectives on classical fields, while maintaining Springer's high standards of academic excellence. Also considered for publication are lecture materials, proceedings, and other related materials of exceptionally high quality and interest. The subject matter should be original and timely, reporting the latest research and developments in all areas of electrical engineering.

The audience for the books in LNEE consists of advanced level students, researchers, and industry professionals working at the forefront of their fields. Much like Springer's other Lecture Notes series, LNEE will be distributed through Springer's print and electronic publishing channels.

More information about this series at http://www.springer.com/series/7818

Andrzej Kłos

Mathematical Models of Electrical Network Systems

Theory and Applications - An Introduction

Springer

Andrzej Kłos
Electrical Department
Warsaw University of Technology
Warsaw
Poland

ISSN 1876-1100 ISSN 1876-1119 (electronic)
Lecture Notes in Electrical Engineering
ISBN 978-3-319-52176-3 ISBN 978-3-319-52178-7 (eBook)
DOI 10.1007/978-3-319-52178-7

Library of Congress Control Number: 2016963175

Printed on acid-free paper

This Springer imprint is published by Springer Nature
The registered company is Springer International Publishing AG
The registered company address is: Gewerbestrasse 11, 6330 Cham, Switzerland

Preface

The electricity and electrical networks are used now in so many fields of science and engineering practice that the further development of theoretical background and practical applications of electrical networks is necessary. The recently used network (circuits) theory and methodology of linear network system analysis and solution are based, mainly, on rather simple mathematical models. The majority of textbooks, handbooks, and lecturing materials concerning electrical network systems, which are used in a high- and medium-level technological universities and technical schools, are outdated. Nowadays the electricity supply utilities have serious problems concerning security, economy, ecology, etc., e.g. power systems have serious problems with development of high voltage networks and building new power stations. This book is needed which is widening the mathematical background (network theory), not so much as to be difficult to the readers, but useful as a tool for the new practical applications. This book presents a modern and non-conventional network theory and its practical applications in network analysis and solution. In the first part of this book, an advanced mathematical (linear algebra) approach to the modeling of time-constant networks is given. The algebraic model of network system topology is defined, and topological equations are derived and expressed in the form of a linear space. It was shown that modeling network graph in terms of linear algebra leads to the non-singular topological transformation matrix *T,* which appears to be a useful tool of network analysis and solution. The algebraic models of the Kirchhoff's current and voltage laws and of the Ohm's law are derived and expressed in the form of linear spaces. It makes possible the derivation of various, commonly not known, equations, which are widening the methodology of network analysis. The combined current–voltage vector is defined, and it leads to the unexpected result; it was proved that the summation of current and voltage values is reasonable from the mathematical viewpoint. Using the algebraic models of currents and voltages, the generally not known mathematical formulations of fundamental Kirchhoff's laws are derived and discussed. The classical Ohm's law is supplemented by introducing the system parameters, which enable using the current and voltage sources in network system.

In the second part of this book, using the algebraic network model, the various applications of this model are presented. The connection between theory and practical network problems is shown by solving the selected examples of network problems in which the variety of input data, solvability conditions, not known solution method, computation efficiency etc., are to be taken into account. The general algebraic model of network solution (generalization of classical method) is presented which may be applied to variety of technical and non-technical fields. Particularly, some examples of practical applications in the field of power system network analysis and solution are given. The usefulness of the new formulations of Kirchhoff's and Ohm's laws is shown using simple examples of network. The solution method of the arbitrary input data problems is given using algebraic network model. Using topological matrix T, the equations of network system analysis and solution are derived and discussed. The not known solution methods of load flow in power system network are derived and illustrated using the example of real network.

The text of this book includes the mathematical derivations and formulas, but it is understandable for engineers and students. Mathematically, more difficult parts, e.g., linear space terminology, are illustrated and described in a way understandable for non-mathematicians. The book level and contents are addressed to researchers, university lecturers, software developers, and advanced undergraduate and post-graduate students involved in power system network analysis and network development.

Warsaw, Poland Andrzej Kłos

Contents

Part I
Mathematical Model of Network System

Chapter 1
Introduction

1.1 Network System

The electricity is an important and widely used product of the today's civilization. The electrical power technology, electronics, information technology, and various kinds of electronic utilities are based on the application of electrical networks.

The term "electrical system" has been used to describe a vast range of physical or abstract systems which differ in appearance, function, and size. In this book, *the electrical network system*, called simply *network system,* is concerned. It has following characteristic features:

1. Network systems are concerned with the generation, transportation, transformation, and utilization of medium what is called generally "electricity." Physical medium is the electric charge, identified sometimes with electrical power and energy.
2. Network systems involve a number of interconnected routes, along which the medium is generated, transported, transformed and utilized, and involve the geographically distributed points, which interconnect these routes. This structure is visualized geometrically by a graph whose branches represent the routes and whose nodes represent interconnection points.
3. The routes are called the network system branches, and the points are called the nodes or the terminals. The network branches represent parts of the network system which are technological subsystems, earth, and air.
4. The network system can be physically described and mathematically modeled by following quantities: current, voltage, electrical energy, electrical power, and electrical parameter. These physical quantities are interrelated with each other and with the topological structure of network graph. The basic interrelations are called the network laws.

A. Kłos, *Mathematical Models of Electrical Network Systems*,
Lecture Notes in Electrical Engineering 412,
DOI 10.1007/978-3-319-52178-7_1

5. The theory of networks is still evolving basing on the matrix theory. The main aim of this book was to widen the theoretical background of network system knowledge and showing the examples of possible practical application of this knowledge.

1.2 Modeling Network System

The term "*mathematical modeling*" is often used in physics and other fields to describe various objects, events, and actions. The term, *mathematical modeling of electrical network system*, used in this book means the description of the physical states and working of an electrical network system, using mathematical notions and expressions to solve the theoretical and practical network problems.

The main step in the formulation of the mathematical models of network systems is to replace the real network by a physical model of the network, which possesses the major characteristics of the original network but in abstract form. Such an abstract form is the starting point to the mathematical model of network. It is assumed that the reader has the knowledge and ability to formulate the abstract model of electrical network. This book starts from such model, which is referred hereafter as the mathematical model of network system. The mathematical model of network system consists of two sets as follows:

- The set of mathematical quantities representing and describing the network system. Each of these quantities represents (is modeling) a physical network quantity and consist of the given data concerning the topology of network, voltages, currents, powers, and branch parameters.
- The set of relations between mathematical quantities representing the physical network laws, and other relations, all expressed in the form of equations, inequalities, and logical expressions.

Network system models are classified into different types depending on the kind of data and assumed relationships, using the following categories:

(a) A network system model is *deterministic* if the data and assumed relationships are known to be correct with no significant errors. The model is *non-deterministic* if there are some degrees of uncertainty associated with some of data or relationships. A non-deterministic model is called *probabilistic* if the data are treated as random variables with certain probability distributions.
(b) A network system model is *linear* if the relations between mathematical quantities are linear, and *nonlinear* if among them there are nonlinear relations. The relations in nonlinear model may be continuous or may be step functions.
(c) A network system model is *dynamic* or *time-independent* if the mathematical quantities are of dynamic character or are fixed in time, and *dynamic* or *time-dependent* if they are of dynamic character or varies in time.

The above classification subdivides the models of network system into eight main classes according to the eight subsets of properties formed by the two major choices in each of (a), (b), and (c) classifications. This is a general classification which does not include models for some special kind of problems, such as those which are mixed with respect to this classification. For example, the important technique of "state estimation" is concerned with finding a "best possible" approximate solution for networks where the data may be incomplete, inaccurate, and contradictory.

In each of the above main classes, the problem can be further subclassified according to the amount and kind of information (data) available. For each of the data categories, topology data, state of network data, and current–voltage relation data, the given data may include either complete information or incomplete, partial information.

In what follows the basic linear and continuous mathematical model of the interconnected, closed electrical network system (called in what follows the mathematical model of network system) is derived. Such model can be used as a part in the modeling of any other kind of network models.

1.3 Solving Network System

It is quite usual to encounter the problem of "solving a network system," though this may mean many different things. In this book, network solution problems of general form are considered. They are classified in manner most useful for the solution process as follows.

Suppose certain information about a network is given, or is assumed, which is of a quantitative nature, and suppose that certain quantitative information about the network is required. If the given information (input data) does not include the required information, or if the information is given but is ambiguous or incorrect, then arises the obvious problem of finding or estimating the required information. It is assumed here that the required information can be found by calculations using the given data rather than by other methods, such as measuring the quantities. In the following, the term *solution of the electrical network system* is used to denote both the set of calculated values of the required information and also the process of calculation involved.

The definition given here for "solution of the electrical network" is very general, since it allows many possible pairings of "information given" with "information required." In order to keep the situation reasonable, it is further assumed that the information required about a network is complete, or nearly complete. More specifically, it is supposed that the information is required on each of the following which is complete as possible:

(a) *The network topology*. The number and the interconnections of all branches and the subdivision of branches into a tree and a cotree. This is equivalent to knowing the network graph.

(b) *The electrical state of network.* The numerical values of all branch currents and branch voltages, or in other words, complete knowledge of the network currents, voltages, and other required information.
(c) *The interrelations of current and voltages* (numerical values of all branch parameters), or in other words the admittance or impedance branch coefficients and the values of voltage and current sources.

The network system is regarded as solved if the above information is known, and the term of "solving the network" is that of finding the above information when it is not part of known data.

1.4 Network Laws

The values of branch currents and branch voltages in the network system are restricted in the sense that the states of network which are physically possible must fulfill the fundamental network laws.

The Kirchhoff's Current law: The sum of branch currents in a node (cut-set), taken into account the directions of branch currents is equal to zero. Note that a current directed away from a node is considered to be the same that the current directed towards the node which has the same numerical value but opposite sign. The law, stated in simple terms means that the current flowing into the node (cut-set) equals the current flowing out of it.

The Kirchhoff's Voltage law: The sum of all branch voltages in a loop-(loop-set) taken into account the directions of branch voltages is equal to zero. This law stated in simple terms, mean that the voltage value between any two nodes of network is independent of the loop (path of branches) between these nodes along which the voltage is measured.

The Ohm's law: Generally, it is certain linear function between the currents and the voltages of network branches. This law has a non-topological nature and is reflecting the interaction in a physical network between the currents/voltages and the physical structure of the branches of the network. The interaction of Ohm's law is done using constant branch coefficients of current/voltage functions. Note that each real network, called in this book *the network system*, must include at least one current or voltage source. Without any source, the network is empty ($I = 0$, $V = 0$) and the classical network Ohm's law in such network can be used to the parts of network only. The Ohms law of network system, derived in this book, refers to the network system.

The Power law: The sum of powers in all branches of network is zero, where the power in a branch is defined as a product of the branch current and branch voltage. Note that these values may be positive or negative according to their relation to the branch direction. This is equivalent to saying that each set of all branch voltages written as a vector is orthogonal to any corresponding set of all branch currents, written as a vector (that is the scalar product of two vectors is zero). The law is equivalent to the property that each cut-set of network has an even number of branches in common with each loop-set. This in turn can easily be shown to follow automatically from the definition used here for cut-set and loop-set, so then in fact the power law is a consequence of the current and voltage laws.

Chapter 2
Basic Notions

2.1 Network Graph Notions

Classical modeling of network systems uses the graph theory notions to define the graph of network. The following fundamental notions are used:

The branch (*edge*) is a basic axiomatic notion and is defined as any element, represented by a line segment, with specified two different ends: *terminals plus* (+) and *terminal minus* (−) designating positive branch direction from terminal minus to terminal plus.

The node (*vertex*) is a basic axiomatic notion and is defined as a point designating one or more branch terminals.

The connection or *interconnection* of two or more branches means that they have one common terminal.

The planar graph is a graph, which if drawn in planar plane has not crossing branches.

The tree is defined as any minimal set of interconnected branches, called *tree branches*, which contain all nodes in network system. From the definition of a tree, it follows that it is an open subgraph of a network system graph.

The cotree is defined as a set of branches, which does not belong to any tree. The sets of tree branches and cotree branches are *complementary* if they are disjoint. Cotree is a dual concept, in any sense, to the tree.

2.2 Electrical Network Notions

Graph theory provides an excellent tool for visualizing and developing the model of network graph. However, graph theory is of little use in modeling the physical states of the network system and interrelations of branch currents and branch voltages. Furthermore, even representation of network by a graphical model has

A. Kłos, *Mathematical Models of Electrical Network Systems*,
Lecture Notes in Electrical Engineering 412,
DOI 10.1007/978-3-319-52178-7_2

serious drawbacks from the point of view of development of the mathematical model of network. Hence, graph theory is used as an aid in developing the model. In what follows, the final model presented is algebraic in character and is expressed in terms of linear algebra. Taking into account the above definitions, the basic notions used in mathematical modeling of network system are defined as follows:

The directed electrical branch (called branch in what follows) is a basic notion and is defined as any electrical element (object or earth or air), represented by a line segment, which has two specified end terminals: plus terminal and minus terminal, designating branch direction.

The electrical network system (called network system) is a set of connected branches, which are electrically interconnected and electrically active. Electrical interconnection means that the electrophysical quantities of all branches (e.g., branch currents) are dependent on each other. In other words, in network system, the electrophysical quantity of every branch (e.g., branch current) depends on electrophysical quantities of all other branches, and the system has the ability to be active (possesses electricity source).

Electrical network system is *closed* if each electrical branch is connected at least with two or more other branches and is *open* if there are branches connected with one terminal only to the other branches.

The branch current is a quantity, which characterizes the amount of medium (electric charge) transmitted along the branch in a unit time interval. It can be measured at any one point of branch, assuming that there are no losses or dissipation of the medium in the branch.

The branch voltage is a quantity, which characterizes the difference of electric "pressure" or "stress" being exerted and measured on the branch between plus terminal and minus terminal.

The above quantities are fundamental notions used in the mathematical modeling of electrical systems. In order to derive the algebraic relations precisely, it is necessary to define the algebraic sign of the branch current and branch voltage values in relation to the branch direction. The choice of sign is optional. In what follows, we assume:

Branch current has *plus sign* if current is flowing from branch terminal plus to branch terminal minus. Branch voltage has *plus sign* if the voltage value measured along the branch is increasing in the direction from terminal minus to terminal plus.

The branch parameters are the constant coefficients of branch current as a function of branch voltage, called *branch admittances* or the constant coefficients of branch voltage as a function of branch current, called *branch impedances*, and are the constant coefficients of electricity sources: *the ideal current sources* and *the ideal voltage sources.*

Real electrical branches are often combinations of the above main kinds, e.g., in power system analysis, there are sending (generating) branches, which are combination of ideal voltage sources and impedances and receiving (load-consuming) branches, which are treated as current sources and admittances.

The node-set is defined as a set of branches connected together to one common terminal (node of a graph). *The directed node-set* can be defined as a set of network branches, which connect one node with the rest of network. The branches defining a directed node-set are connected to the node (branch terminal) either by the terminal plus or by the terminal minus. So there are two options and consequently two possible definitions of the *direction* of node-set. In the first option, the direction of node-set is the same as the direction of branches directed toward common terminal, and in the second option, it is the same as the direction of branches directed away from the common terminal. From the definition of the node-set, it follows that the sum of all node-sets is equal to zero. This follows because each branch is represented in one node-set by plus terminal and in another node-set by minus terminal. Removing any one node-set in the network must leave a sum of the remaining node-sets not equal to zero. The number of all node-sets minus any one is denoted by n and is a number of linearly independent node-sets.

The cut-set is defined as a set of network branches connecting two parts of network. It is the generalization of node-set. Note that a node-set is also connecting two parts of network, because it connects one node with the rest of network, so is a special kind of cut-set. Graphically, cut-set is a set of branches cut by a line disconnecting two parts of network. From the definition of a cut-set, it follows that the number of cut-sets is equal to the number of all subsets of the $n + 1$ node-sets, excluding the empty subset and the whole subset, so the number of cut-sets is equal to $2^{n+1} - 2$.

The path of branches is intuitively defined as any set of two or more branches and nodes (branch terminals) connected in series (one branch after another). The number of nodes is equal to the number of branches plus one. The path of branches is closed if two ends of path terminals are one terminal. The direction of path is optional and may be defined equal to the direction of any branch in path.

The loop is generally defined as a closed path which is not crossing itself. Intuitively, it is a loop or a mesh in planar networks.

Chapter 3
Algebraic Model of Network Graph

The important aspect of the mathematical modeling of network system is the representation of electrical network graph in terms of mathematical quantities (scalars, vectors, matrices, linear spaces), which represent the physical quantities of the network system graph (branches, node-sets, cut-sets, loop-sets) and algebraic relations representing this mathematical quantities. In this chapter, the electrical network graph is modeled in terms of linear algebra. The physical quantities and relations are modeled using the b-dimensional Euclidean linear space $\overline{A}$, span by the orthonormal basis of vectors a_1, a_2 ... a_b.

3.1 Cut-Set and Loop-Set Vectors

Consider the electrical network graph of b-directed branches. Let us assume that branches are numbered from 1 to b. To each axis of the linear space $\overline{A}$, the electrical network branch j is assigned (see Fig. 3.3).

The branch vector, denoted by b_j (index j is a number of branch), is modeled as a b-dimensional vector with one nonzero element equal to plus one in place of branch number, e.g., a branch vector number 3 is written as follows:

$$b_3 = [0_1\, 0_2\, 1_3 \ldots .0_{b-1}.0_b]^{\mathrm{T}} \tag{3.1}$$

where 1_3 denotes a real network branch number $j = 3$.

The cut-set vector C_s is defined generally as a b-dimensional vector, which is the sum of branch vectors, which form the directed cut-set.

$$C_s = \sum b_c \tag{3.2}$$

where b_c are the branch vectors belonging to the cut-set vector.

A. Kłos, *Mathematical Models of Electrical Network Systems*,
Lecture Notes in Electrical Engineering 412,
DOI 10.1007/978-3-319-52178-7_3

The node-set vector is a special kind of a cut-set vector, which is a sum of branch vectors, which form the directed node-set.

The set of linearly independent node-set vectors is a set of all minus any one node-set vector in the network. In what follows, the node-set vectors are identified with the cut-set vectors. In what follows, any set of cut-sets may include the node-sets and the cut-sets.

The loop-set is defined in Chap. 2 as a set of branches forming a closed path.

The loop-set vector L_s can be intuitively defined as a b-dimensional vector, which is the sum of branch vectors forming the directed loop-set.

$$L_s = \sum b_l \tag{3.3}$$

where b_l are branch vectors belonging to the loop-set vector.

However, the loop-set vectors are in a certain sense a dual concept to that of the cut-set vectors. Consequently, in terms of linear algebra, the *loop-set vector* is a b-dimensional vector defined as follows:

- Its components are all equal to +1 or −1 or 0;
- It is orthogonal to every cut-set vector.

So the definition (3.3) is interpreted as follows: Any vector of components 1, −1, 0, whose inner product (scalar product) with every cut-set vector is zero, is the loop-set vector.

3.2 Topology of Network

The network analysis and solution are based on finding the linearly independent loop-sets and cut-sets. The loop-sets and cut-sets are fundamental notions in the mathematical modeling of network systems, because they are used in formulations of the Kirchhoff's laws. In particular, a set of m linearly independent loop-set vectors and a set of n linearly independent cut-set vectors are used in formulations of the current and voltage relations. Taking into account the great number of possible cut-sets and loop-sets in real networks, the finding of linearly independent sets may be very difficult. Both sets can be easily found after defining network topology. The term "topology" relates to the graph of network but may have various meanings. In what follows, let us define the topology of electrical network as follows:

The topology of electrical network is defined as a subdivision of the b-branch network graph into the tree (any set of n tree branches) and the complementary cotree (a set of m complementary (not belonging to tree) branches.

$$n + m = b. \tag{3.4}$$

Generally, there are many topological tree–cotree subdivisions of a given network graph. The choice of any subdivision means that the network is topologically oriented. The algebraic operations, using the b-dimensional vectors and matrices, require the ordering of branches to be appropriate to the chosen topology. Generally, the ordering is not limited. Taking into account the ordering which is often used in practice, we assume that the first m rows and columns of vectors and matrices are associated with the cotree branches and the last n rows and columns of vectors and matrices are associated with the tree branches.

The vector of branch numbers, denoted by B_j, is subdivided into two subvectors:

$$B_j = \begin{bmatrix} B_m & B_n \end{bmatrix} \tag{3.5}$$

where

B_m vector of units, of cotree branches.
B_n vector of units, of tree branches.

In the next sections, the topological models of loop-sets and cut-sets are derived.

3.3 Topological Model of Loop-Sets

Each cotree branch generates a unique loop-set. As it follows from the definition of a tree/cotree topology, the tree is a subgraph, which includes all network nodes. Consequently, every cotree branch connects two nodes, each of them belonging to a tree. The tree is open subgraph; so for every cotree branch, there must exist a set of tree branches which, together with this cotree branch, form a closed path in the network.

The topological loop-set vector generated by a cotree branch (called in what follows *cotree loop-set vector* or *loop-set vector*), denoted L_c, is defined as the following sum of branch vectors: one cotree branch vector and the unique set of tree branch vectors, which form together a closed path in the network:

$$L_c = \sum b_c \tag{3.6}$$

where b_c are branch vectors belonging to the cotree loop-set vector.

Direction (arrow) of a cotree branch, which generates the loop-set vector, defines the direction of this loop-set vector. So all the tree branches belonging to loop-set, if they have the same direction as the direction of co-tree branch which generate the loop-set, they have +1 in loop-set and if they have the opposite direction they have −1 in the loop-set (see Fig. 3.1).

Every cotree branch generates a loop-set vector, so in any topological structure of network, there are m topological loop-set vectors.

The topological loop-set matrix (called in what follows loop-set matrix), denoted L, represents a set of m linearly independent topological loop-set vectors. Each row of matrix L is the topological loop-set vector.

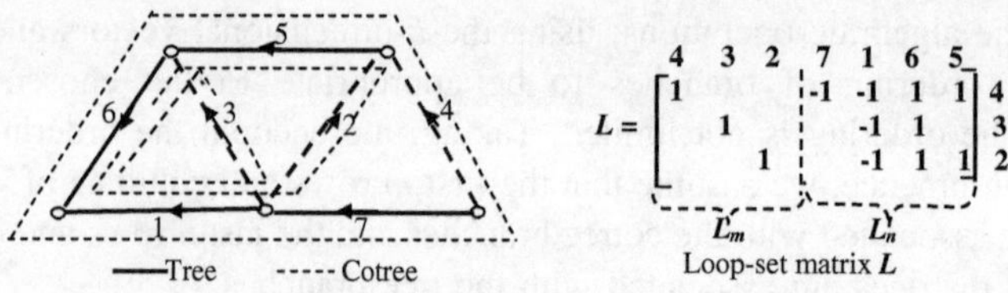

Fig. 3.1 Example of loop-sets and loop-set matrix L

$$L = [L_{c,1} \quad L_{c,2} \quad L_{c,3} \quad . \quad L_{c,k} \quad . \quad L_{c,m}]^{\mathrm{T}} \tag{3.7}$$

where $L_{c,k}$ are the loop-set vectors (rows of matrix L).

The linear independency of m loop-set vectors follows from the definition of loop-set vector. A loop-set vector cannot be any sum of remaining loop-set vectors because each sum must include two or more cotree branches what contradicts with the definition of loop-set vector. Loop-set matrix is of order $m \times b$ and of rank m. Generally, any loop-set incidence matrix can be subdivided after appropriate rearrangement of columns, into two submatrices:

$$L = [L_m \quad L_n] \tag{3.8}$$

where L_m is a non-singular matrix of order $m \times m$ and rank m and L_n is of order $m \times n$.

Taking into account the topological ordering of branches (see Eq. 3.5) and that the loop-sets are generated by cotree branches, the particularly useful form of loop-set matrix L is as follows (see Fig. 3.1):

$$L = [1^m \quad L_n] \tag{3.9}$$

where

L_n is the tree–cotree incidence matrix of order $m \times n$.
1^m is a unit matrix of order $m \times m$ of m cotree branches.

The matrix L_n is a well-known incidence matrix relating tree branches to cotree branches and is used frequently in the network analysis and solution.

3.4 Topological Model of Cut-Sets

The tree is any open set of branches in network. Each tree branches generate a cut-set generate a cut-set. The cut-set is a set of branches including one set of cotree branches which together fulfill the current law. Visually, the cut-set can be illustrated on a graph of network by dotted line (not full line) which is cutting the cur-set branches.

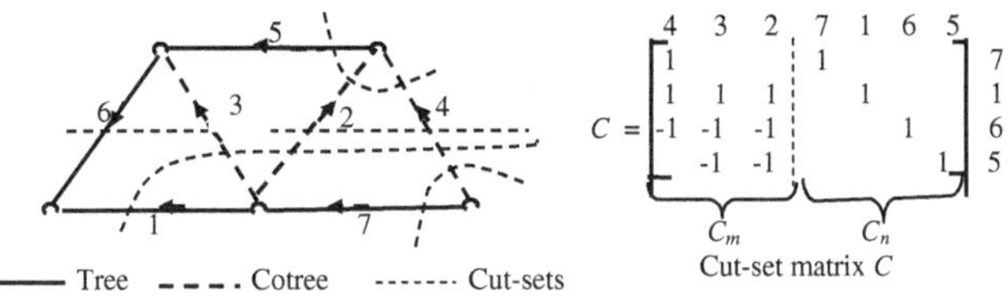

Fig. 3.2 Example of the cut-sets and cut-set matrix C

E.g., in Fig. 3.2 the cut-set generated by tree-branch 1 is cut (include) the branches 1, 2, 3, and 4. Algebraically such set of branches is used in form of a cut-set vector, and a set of all cut-sets in network is used, in form of a cut-set matrix (see Fig. 3.2).

The cut-set vector generated by a tree branch (in what follows called *tree cut-set vector* or *cut-set vector*), denoted C_t, is defined as a sum of the following branch vectors: one tree branch vector and a set of cotree branch vectors subdividing the network into two subnetworks.

$$C_t = \sum b_t \tag{3.10}$$

where b_t are the branch vectors belonging to the tree cut-set vector.

Direction (arrow) of a tree branch that generate the cut-set defines the direction of cut-set vector. So the all cotree branches belonging into cut-set if they have the same direction as direction of the tree branches then have number +1 in cut-sector; if they have opposite direction then have number −1 in cut-set vector (see Fig. 3.2) Each tree branch generates the cut-set vector, so in any network there are an independent cut-set vectors.

The cut-set matrix, denoted C, represents a set of n linearly independent cut-set vectors.

$$C = \begin{bmatrix} C_{t,1} & C_{t,2} & C_{t,3} & \cdots & C_{t,k} & \cdots & C_{t,n} \end{bmatrix}^{\mathrm{T}} \tag{3.11}$$

where each $C_{t,k}$ is a cut-set vector (row of matrix C).

The linear independency of n cut-set vectors follows from the definition of cut-set. A cut-set vector cannot be any sum of remaining cut-set vectors because each sum must include two or more tree branches what contradicts with the definition of cut-set. Cut-set matrix is of order $n \times b$ and of rank n. Generally, any cut-set incidence matrix can be partitioned, after appropriate reordering of columns, into two submatrices:

$$C = [C_m | C_n] \tag{3.12}$$

where C_n is a non-singular matrix of order $n \times n$ and rank n; C_m is of order $n \times m$.

Taking into account the topological ordering of branches has (see Eq. 3.5), the branches associated with columns of cut-sets matrix C are ordered as follows: first m branches are the cotree branches and last n are the tree branches. So the matrix C is partitioned as follows:

$$C = [C_m | 1^n] \tag{3.13}$$

where

C_m *the tree incidence matrix* of order $n \times m$ and
1^n is a unit matrix of order $n \times n$ representing tree branches.

Matrix C is illustrated in Fig. 3.2. The matrix C_m is a well-known incidence matrix relating cotree branches to the tree branches and is used frequently in the network analysis and solution.

3.5 Orthogonality of Cut-Sets and Loop-Sets

From the definition of loop-set vectors (3.6) and cut-set vectors (3.10), it follows that they are orthogonal to each other.

$$C_t L_c^{\mathrm{T}} = L_c C_t^{\mathrm{T}} = 0 \tag{3.14a}$$

It means that the loop-set matrices (3.8 and 3.9) and cut-set matrices (3.12 and 3.13) are orthogonal to each other.

$$CL^{\mathrm{T}} = LC^{\mathrm{T}} = 0 \tag{3.14b}$$

The orthogonality of L and C reflects what is called the duality of network topology, which is an important feature of electrical networks. The various practically useful interrelationships can be derived using orthogonality of cut-sets and loop-set matrices. Equations (3.14a, b) make possible finding matrix C if matrix L is known or finding matrix L if matrix C is known. Such relations can be derived as follows:

Using orthogonal relation of matrix C in the form $C = [C_m \ \ C_n]$ and matrix L in the form $L = [\,1^m \quad L_n\,]$.

$$[C_m \ C_n] \begin{bmatrix} 1^m \\ L_n^{\mathrm{T}} \end{bmatrix} = 0 \tag{3.15}$$

$$C_m\, 1^m + C_n L_n^{\mathrm{T}} = 0$$

$$L_n^{\mathrm{T}} = -C_n^{-1}\, C_m$$

Hence,

$$L = \left[\, 1^m \quad (-C_n^{-1}\, C_m)^{\mathrm{T}} \,\right] \tag{3.16}$$

what means that matrix L can be found using matrix C

Using orthogonal relation of matrix C in the form $C = [C_m\ 1^n]$ and $L = [\,L_m \quad L_n\,]$.

$$[L_m\ L_n] \begin{bmatrix} C_m^{\mathrm{T}} \\ 1^n \end{bmatrix} = 0 \tag{3.17}$$

$$L_m\, C_m^{\mathrm{T}} + L_n\, 1^n \ = 0$$

$$C_m^{\mathrm{T}} = -L_m^{-1}\, L_n$$

Hence,

$$C = \left[\, (-L_m^{-1}\, L_n)^{\mathrm{T}} \quad 1^n \,\right] \tag{3.18}$$

What means that matrix C can be found using matrix L

Using orthogonal relation of matrices $C = [\, C_m \quad 1^n \,]$ and $L = [\, 1^m \quad L_n \,]$.

$$[C_m\ 1^n] \begin{bmatrix} 1^m \\ L_n^{\mathrm{T}} \end{bmatrix} = 0 \tag{3.19}$$

$$C_m\, 1^n + 1^m\, L_n^{\mathrm{T}} = 0$$

Hence,

$$C_m \ = -L_n^{\mathrm{T}} \tag{3.20}$$

$$L_n \ = \ -C_m^{\mathrm{T}} \tag{3.21}$$

The relations (3.20 and 3.21) are very useful in practical applications.

3.6 Linear Space Model of Network Topology

From the engineering point of view, it would be of some advantage to visualize the topological models, described above, in a linear space $\overline{A}$. Visual illustration of linear

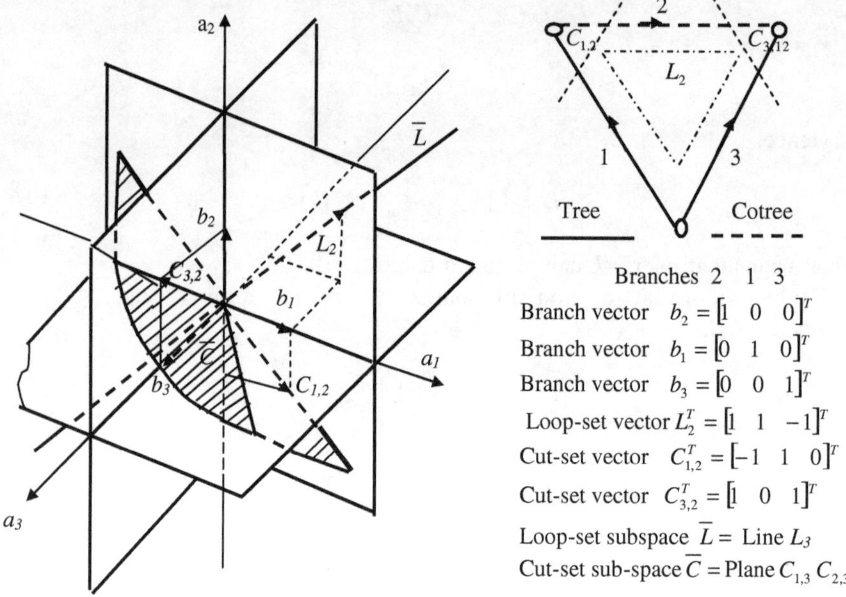

Fig. 3.3 The three-dimensional linear space illustration of subspaces $\overline{L}$ and $\overline{C}$

space model of network topology can be done for very simple, 3-branch network only (see Fig. 3.3). In order to visualize the network topology, in terms of linear algebra, the loop-set and cut-set vectors (the rows of the matrices L and C) are transposed into column matrices L^{T} and C^{T}. Generally, the loop-set and cut-set vectors and matrices can be modeled as a linear subspaces of space $\overline{A}$ (see Fig. 3.3). The set of m loop-set vectors L (in Fig. 3.3 one vector L_3) form the *loop-set subspace* $\overline{L}$ of space $\overline{A}$. Analogically, the set of n cut-set vectors (in Fig. 3.3 vector C_{v1} and C_{v2}) form the linear *cut-set subspace* $\overline{C}$ of space $\overline{A}$. The subspaces $\overline{L}$ and $\overline{C}$ are orthogonal to each other. They are graphically illustrated in Fig. 3.3 for a simple example of 3-branch network.

3.7 Topological Transformation

The above-derived linear space model leads to a new algebraic structure. The cut-set subspace and the loop-set subspace are orthogonal to each other. The m loop-set vectors (matrix L^{T}) and the n cut-set vectors (matrix C^{T}) are, so far, vectors in space $\overline{A}$ (let us call it the **old** space). Note that the loop-set vectors and the cut-set vectors span a **new** b-dimensional space (let us call it the **new** space (see Fig. 3.3). Consequently, in the **new** space, the loop-set and cut-set vectors are the basis of **new** space, so they are unit vectors in this space. It means that in the **old** space, the loop-set matrix $L^{\mathrm{T}} = \begin{bmatrix} 1^m \\ L_n^{\mathrm{T}} \end{bmatrix}$ and the cut-set matrix $C^{\mathrm{T}} = \begin{bmatrix} C_m^{\mathrm{T}} \\ 1^n \end{bmatrix}$ are modeled by a joint matrix:

$$\begin{bmatrix} L^{\mathrm{T}} & C^{\mathrm{T}} \end{bmatrix}_{\mathrm{old}} = \begin{bmatrix} 1^m & C_m^{\mathrm{T}} \\ L_n^{\mathrm{T}} & 1^n \end{bmatrix}. \tag{3.22}$$

In the **new** space, the loop-set and cut-set matrices are modeled by the joint unit matrix:

$$\begin{bmatrix} L^{\mathrm{T}} & C^{\mathrm{T}} \end{bmatrix}_{\mathrm{new}} = \begin{bmatrix} 1^m & 0 \\ 0 & 1^n \end{bmatrix} \tag{3.23}$$

Taking into account Eqs. (3.22 and 3.23), the transformation equation of loop-sets and cut-sets from **new** to **old** space is as follows:

$$\begin{bmatrix} 1^m & 0 \\ 0 & 1^n \end{bmatrix} \begin{bmatrix} 1^m & C_m^{\mathrm{T}} \\ L_n^{\mathrm{T}} & 1^n \end{bmatrix} = \begin{bmatrix} 1^m & C_m^{\mathrm{T}} \\ L_n^{\mathrm{T}} & 1^n \end{bmatrix}$$

It means that the transformation matrix of loop-sets and cut-sets from new to old space is equal to $\begin{bmatrix} L^{\mathrm{T}} & C^{\mathrm{T}} \end{bmatrix}$.

Note that interpreting physically, the above transformation means transformation of m cotree branches into m loop-sets, which are defined by this m cotree branches; and means transformation of n tree branches into n cut-sets, which are defined by this n tree branches (see Sects. 3.3 and 3.4).

The following definitions are to be formulated:

The **new** space is *the topological space.* It is the b-dimensional linear space, denoted $\overline{T}$, in which unit basis vectors indicate the cotree and tree topological structure of network system.

If transformed to space, $\overline{A}$ (old space) is equal to matrix $\begin{bmatrix} L^{\mathrm{T}} & C^{\mathrm{T}} \end{bmatrix}$.

The topological transformation matrix denoted T (in what follows called *topological matrix T* or *matrix T)* is transformation matrix of vectors in space $\overline{T}$ into vectors in space $\overline{A}$.

$$T = \begin{bmatrix} L^{\mathrm{T}} & C^{\mathrm{T}} \end{bmatrix} = \begin{bmatrix} 1 & C_m^{\mathrm{T}} \\ L_n^{\mathrm{T}} & 1 \end{bmatrix} \tag{3.24}$$

The topological transformation matrix T is an important quantity, which makes possible widening the analysis of network systems. It will be often used in the next chapters.

3.8 Topological Matrix T

The topological transformation matrix T takes the simplest and practically useful forms, based on the well-known incidence matrices C_m^{T} and L_n^{T}.

Substituting Eqs. (3.20–3.24)

$$T = \begin{bmatrix} 1 & C_m^{\mathrm{T}} \\ -C_m & 1 \end{bmatrix} \tag{3.25}$$

and substituting Eqs. (3.21–3.24)

$$T = \begin{bmatrix} 1 & -L_n \\ L_n^t & 1 \end{bmatrix} \tag{3.26}$$

The topological matrix T has some interesting properties:

- For a given network, there are as many matrices T as there are pairs of: n independent cut-sets (nodes) and m independent loop-sets (meshes).
- All matrices T are similar and have the same nonzero integer determinant. Note that it comes from the linear independency and orthogonality of incidence matrices C and L.
- The matrix T is non-singular because the columns of L^T and columns of C^{T} are linearly independent and orthogonal to each other.
- The matrix T^{-1} is not a (0, 1, −1) incidence matrix. The values of its entries are greater than −1 and are less than 1.
- The matrix $T^{\mathrm{T}}T$ describes the structure of loop-sets and cut-sets, e.g., the lower right submatrix of a matrix $T^{\mathrm{T}}T$ shows the structure of cut-sets as follows: Each row shows the relation of a cut-set generated by tree branch j to the remaining cut-sets. The diagonal element of row is the numbers of branches belonging to the cut-set generated by tree branch j. The off-diagonal elements are the numbers of branches, belonging to the remaining cut-sets, which are common with branches of cut-set generated by branch j. The plus/minus sign of the off-diagonal elements denotes the same (plus) or opposite (minus) direction of branch in cut-set j and the directions of branch in remaining cut-sets. The direction of a cut-set is defined by the direction of tree branch generating this cut-set.
- Analogically the upper left submatrix of $T^{\mathrm{T}}T$ describes the structure of a set of loop-sets.

Figure 3.4 illustrates the matrices T, T^{-1}, and $T^{\mathrm{T}}T$ for a simple example of network.

Matrix T is a powerful tool enabling the widening of electrical network analysis and methodology of network solution. The algebraic model of network developed in the next chapters is based on the application of matrix T. Note that in the

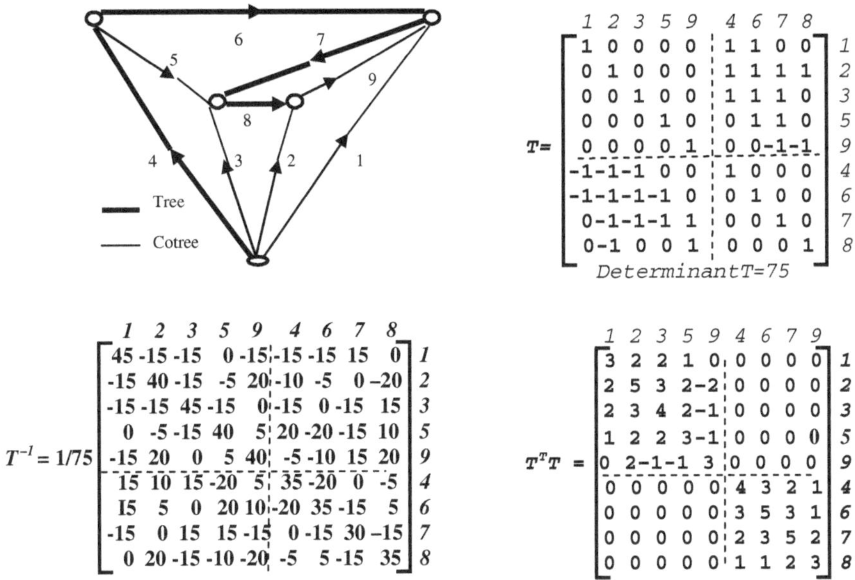

Fig. 3.4 Topological matrices T, T^{-1}, and $T^{\mathrm{T}}T$ for a simple example of network

above-presented algebraic model of network, the primary notion (not to be mathematically defined) is the network branch only. Different than that in graph theory, the node is defined as a set of branches. The application of the above formulas does not need introducing the notion of slack node.

Chapter 4
Algebraic Model of Network Currents

4.1 Matrix Model of Network Currents

The linear algebra model of network currents can be derived starting from the model of branch current. Consider the network formed from b electrical branches, numbered from 1 to b. To every branch, the physical quantity, called the *branch current,* denoted i_j (j is the branch number) is associated, which may have a real or complex value. The branch current i_j has positive sign if its direction of flaw is the same as the direction of a network branch and has negative sign otherwise.

The current vector of network denoted I is defined as a column vector of branch currents.

$$I = \begin{bmatrix} i_1 & i_2 & i_3 & \ldots & i_b \end{bmatrix}^{\mathrm{T}} \tag{4.1}$$

The elements of current vector satisfy fundamental Kirchhoff's current law which ensures that for every cut-set the sum of currents, over all branches belonging to this cut-set, is equal to zero. If the ordering of branches is not restricted, then the current law is expressed in matrix form as follows:

$$C_{\mathrm{s}} I = 0 \tag{4.2}$$

where C_{s} is any possible row cut-set vector, which has the same branch ordering then vector I.

If the network topology (tree/cotree) is defined, not loosing generality, then we assume that the set of n cut-sets, generated by a tree, are the rows of a cut-set matrix, denoted in what follows by C. The ordering of columns in this matrix is as follows: The first m columns are associated with cotree branches and the last n with tree branches. Using such topology, the current law is expressed as follows (see Chap. 3, Sect. 3.4):

A. Kłos, *Mathematical Models of Electrical Network Systems*,
Lecture Notes in Electrical Engineering 412,
DOI 10.1007/978-3-319-52178-7_4

$$C\,I = 0 \tag{4.3}$$

where $C = [\,\mathrm{C}_m \quad 1^n\,]$ is *cut-set matrix* generated by a tree.

The ordering of branches in vector I is as follows:

$$I = \begin{bmatrix} I_m \\ I_n \end{bmatrix} \tag{4.4}$$

where

I_m is *the current state vector* of order m. It is the current vector of cotree branches
I_n is the current vector of tree branches

Vector I_n can be found using current state vector from Eq. (4.3) written as follows.

$$[\,C_m \quad 1^n\,]\begin{bmatrix} I_m \\ I_n \end{bmatrix} = C_m I_m + 1^n I_n = 0 \tag{4.5}$$

where C_m is the *tree–cotree incidence matrix* of order $n \times m$

Solving Eq. (4.5) for I_n gives

$$I_n = -C_m I_m \tag{4.6}$$

The current of tree branches I_n can be also found, using loop-set matrix L. Substituting C_m from Eq. (3.20) into (4.6), we have:

$$I_n = L_n^{\mathrm{T}} I_m \tag{4.7}$$

Finally substituting Eq. (4.7) into (4.4), we have:

$$I = L^{\mathrm{T}} I_m \tag{4.8}$$

The current vector of cotree branches I_m is *the current state vector*. Knowing vector I_m, the current vector of tree branches can be found from Eqs. (4.6) and (4.7), called the *current state equations*. Equation (4.8) is the particular expression of current law Equation (4.3). Note the following difference between the formulations of the above equations: the current law (4.3) and current state Equation (4.6) are expressed using the tree cut-sets, but the same law in Eq. (4.8) and current state Equation (4.7) is expressed using the cotree loop-sets.

4.2 Linear Space Model of Network Currents

In this section, the currents of network system are expressed using the linear algebra model. The algebraic model of the current relationships is derived using the linear space model of network topology described in Sect. 3.6.

Let us model the currents of a b-branch electrical network using the b-dimensional linear space $\bar{A}$. The set of all b-dimensional branch current vectors I satisfying Eq. (4.3) for any cut-set matrix C is a linear subspace of the space $\bar{A}$ and is called the *current space* $\bar{I}$. From Eq. (4.3), it follows that the current subspace $\bar{I}$ is orthogonal to the cut-set subspace $\overline{C}$, and the subspace $\overline{C}$ is orthogonal to the loop-set subspace $\overline{L}$. It means that subspace $\bar{I}$ coincides with the loop-set subspace $\overline{L}$. Taking into account this coincidence, the subspace $\bar{I}$ is spanned by the basis of m loop-set vectors L (see Sect. 3.6).

The linear space model of network currents can be illustrated by a very simple example of 3-branch network in Fig. 4.1. Following topology of network is chosen: Branches 1 and 3 are the tree and branch 2 is a cotree. The cut-set vectors defined by $m = 2$ tree branches are cut-set vector C_1 (generated by tree branch 1) and cut-set vector C_3 (generated by tree branch 3), which span the two-dimensional cut-set subspace $\overline{C}$ (plane). The loop-set vector defined by $n = 1$ cotree branch is loop-set vector L_2 (generated by cotree branch 2) which spans the one-dimensional loop-set subspace $\overline{L}$ Assuming the values of branch currents as $i_1 = i_2 = 1$ and $i_3 = -1$, the current vector $I = [1 \quad 1 \quad -1]$ spans the current subspace $\bar{I}$ (line). The current subspace $\bar{I}$ is orthogonal (perpendicular) to the cut-set subspace $\overline{C}$.

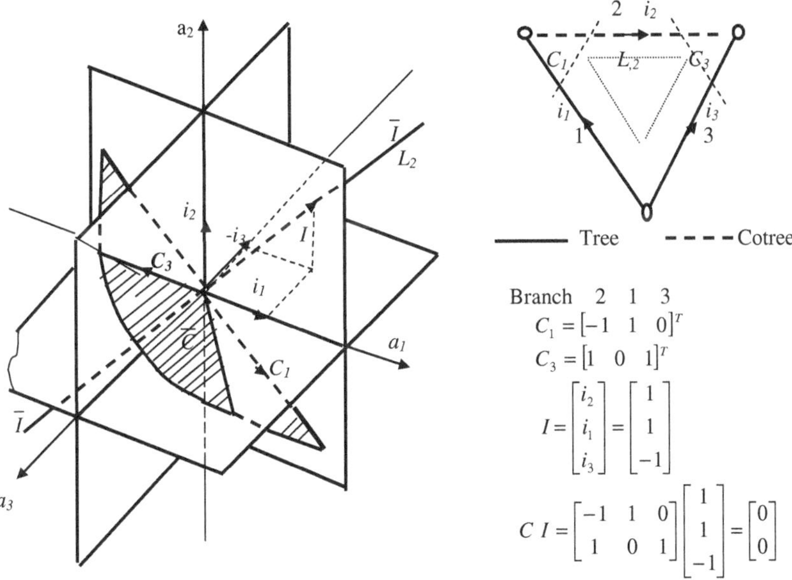

Fig. 4.1 Linear space model of currents for a simple example of 3-branch network

4.3 Topological Transformation of Current Space

So far the currents of network system are modeled in space $\bar{A}$. Taking into account that the cut-set vectors and loop-set vectors span the b-dimensional topological space $\overline{T}$ (see Chap. 3, Sect. 3.7), the network current vector can also be expressed in a new basis, namely in basis of topological space $\overline{T}$. It means that there exists linear transformation of vectors from space $\overline{T}$ to space $\bar{A}$. As it was shown in Sect. 3.7, the topological matrix T is the transformation matrix of vectors between space $\bar{A}$ and topological space $\overline{T}$. In order to find the current vector I (defined in space $\bar{A}$) transformed into space $\overline{T}$, let us write vector I as follows:

$$I = I_{m,o} + I_{o,n} = \begin{bmatrix} I_m \\ 0 \end{bmatrix} + \begin{bmatrix} 0 \\ I_n \end{bmatrix} = \begin{bmatrix} I_m \\ I_n \end{bmatrix} \tag{4.9}$$

where $I_{m,o}$ and $I_{o,n}$ are the cotree and tree branch current vectors of order b.

It can be easily proved that the current vector $I_{m,o}$ after topological transformation from space $\overline{T}$ to space $\bar{A}$ is the current vector I. Starting from Eq. (4.8)—$I = L^{\mathrm{T}} I_m$ and substituting into this equation the cotree current vector I_m of order m by the cotree current vector $I_{m,o}$ of order b, and by substituting the loop-set matrix L^{T} into the matrix of topological transformation $T = \begin{bmatrix} L^{\mathrm{T}} & C^{\mathrm{T}} \end{bmatrix}$, we have:

$$I = \begin{bmatrix} L^{\mathrm{T}} & C^{\mathrm{T}} \end{bmatrix} I_{m,o} \tag{4.10}$$

Note that:

$$\begin{bmatrix} L^{\mathrm{T}} & C^{\mathrm{T}} \end{bmatrix} I_{m,o} = \mathrm{L}^{\mathrm{T}} I_m = I: \tag{4.11}$$

It means that current vector $I_{m,o}$ is vector of topological space $\overline{T}$, and if transformed using matrix T is the current vector I of space $\bar{A}$.

$$I = T\, I_{m,o} \tag{4.12}$$

All vectors $I_{m,o}$ form the subspace $\overline{I}_{m,o}$ of space $\overline{T}$. Both vectors I and $I_{m,o}$ are the same current vector in two spaces: in subspace $\overline{I}$ of space $\bar{A}$ and in subspace $\overline{I}_{m,o}$ of space $\overline{T}$. Note that vector $I_{m,o}$ is b-dimensional, and because of the orthogonality of the spaces $\overline{C}$ and $\overline{I}$, vector $I_{m,o}$ cannot involve any linear combinations of the columns of matrix L, so $I_{m,o}$ has only n nonzero components.

Both the vectors $\mathrm{I}_{m,o}$ and I_m determine the current state of network, and each of them are called the *current state vectors*. They differ by dimension only; $I_{m,o}$ is b-dimensional and I_m is m-dimensional vector.

Illustrating the relation (4.12), in case of three-dimensional current space in Fig. 4.1, let us substitute vector $I = \begin{bmatrix} 1 \\ 1 \\ -1 \end{bmatrix}$, vector $I_{m,o} = \begin{bmatrix} 1 \\ 0 \\ 0 \end{bmatrix}$, and matrix $T = \begin{bmatrix} 1 & -1 & 1 \\ 1 & 1 & 0 \\ -1 & 0 & 1 \end{bmatrix}$.

Equation (4.12) is as follows:

$$\begin{array}{ccccc} & & \begin{matrix} 2 & 1 & 3 \end{matrix} & & \\ \begin{bmatrix} 1 \\ 1 \\ -1 \end{bmatrix} & = & \begin{bmatrix} 1 & -1 & 1 \\ 1 & 1 & 0 \\ -1 & 0 & 1 \end{bmatrix} & \begin{bmatrix} 1 \\ 0 \\ 0 \end{bmatrix} \\ I & = & T & I_{m,o} \end{array} \tag{4.13}$$

4.4 Current Relations Using Topological Matrix *T*

Using the topological transformation matrix *T*, various interrelations between the *b*-dimensional current vectors can be derived. The starting points are Eqs. (4.10) and (4.12).

Matrix *T* is non-singular, so the current state vector $I_{m,o}$ can be found using Eq. (4.12) as follows:

$$I_{m,o} = T^{-1} I \tag{4.14}$$

Interpreting physically the above equation, it means that the set of network currents, which is dependent on each other (must fulfill the current law), can be substituted and considered using the subset of these network currents, which are independent of each other.

A relation between $I_{m,o}$ and $I_{o,n}$ comes as a result of substituting Eq. (4.9) into (4.12).

$$I_{o,n} = (T - 1) I_{m,o} \tag{4.15}$$

This is just a *b*-dimensional version of relation (4.6) or (4.7). Matrix $(T - 1)$ is singular, so if the tree currents are known one cannot find the cotree currents. The substitution of Eq. (4.10) into (4.14) gives the relation between $I_{o,n}$ and *I*:

$$I_{o,n} = (1 - T^{-1}) I \tag{4.16}$$

If the transposed topological matrix T^{T} is used, then it leads to a interesting relation.

$$T^{\mathrm{T}} I = I_m^L \tag{4.17}$$

where I_m^L can be called *the loop current vector*. In order to show that vector I_m^L has physical meaning, note that

$$I_m^L = T^{\mathrm{T}} I = \begin{bmatrix} L \\ C \end{bmatrix} I = LI, \quad \text{because } CI = 0. \tag{4.18}$$

Hence, each entry of I_m^L is the algebraic sum of the branch currents along the loop-set defined by a cotree branch. It means that having m numbers, each of them being the algebraic sum of currents along one of the m loop-sets, one can fined all branch currents in the network. Inverting Eq. (4.17), we have:

$$I = (T^{\mathrm{T}})^{-1} I_m^L \tag{4.19}$$

Substituting Eq. (4.11) into (4.19) gives the relation between $\mathrm{I}_{m,o}$ and I_m^L.

$$I_{m,o} = (T^{\mathrm{T}} T)^{-1} I_m^L \tag{4.20}$$

From Eqs. (4.19) and (4.20), it follows that the current vector I_m^L uniquely defines the current state of network. Substituting Eq. (4.19) into (4.16) gives the relation between I_m^L and $I_{o,n}$.

$$I_{o,n} = (1 - T^{-1})(T^{\mathrm{T}})^{-1} I_m^L \tag{4.21}$$

The above-given relation does not exhaust all possible one. Reducing the relations to the order m and n enables the derivation of various practically useful equations. As it comes from the above relations using the algebraic model of electrical network the non conventional relations can be derived, which may be useful in the network analysis and solution.

Chapter 5
Algebraic Model of Network Voltages

5.1 Matrix Model of Network Voltages

Analogically as in the case of currents, the linear algebra model of network voltages can be derived starting from the model of branch voltage. The algebraic model of network voltage relationships is derived using the linear space model of network topology described in Sect. 3.6. Consider the network formed from b branches. To every electrical branch, the physical quantity, called the *branch voltage* v_j (j is the branch number), is associated, which may have a real or complex value. The branch voltage v_j has positive sign if the direction of the positive voltage drop along the branch is opposite to the direction of a network branch and has negative sign otherwise.

The voltage vector of the network denoted V is defined as a column vector of branch voltages.

$$V = \begin{bmatrix} v_1 & v_2 & v_3 & \ldots & v_b \end{bmatrix}^{\mathrm{T}} \tag{5.1}$$

The elements of voltage vectors satisfy the fundamental Kirchhoff's voltage law, which ensures that for every loop-set, the sum of voltages, over all branches belonging to the loop-set, is equal to zero. The voltage law is expressed in the matrix form as follows:

$$L_{\mathrm{s}}\ V = 0 \tag{5.2}$$

where L_{s} is any possible *row loop-set vector* which has the same branch ordering than vector V.

If the network topology is defined, then, not loosing generality, we assume that the set of m loop-sets, generated by a cotree, are the rows of a loop-set matrix, denoted in what follows by L. The ordering of columns in this matrix is as follows:

A. Kłos, *Mathematical Models of Electrical Network Systems*,
Lecture Notes in Electrical Engineering 412,
DOI 10.1007/978-3-319-52178-7_5

The first m columns are associated with cotree branches and the last n with tree branches. Using such topology, the current law is expressed as follows:

$$L\ V = 0 \tag{5.3}$$

where $L = [\ 1^m\ \ L_n\ ]$ is the loop-set matrix [see Chap. 3, Eq. (3.9)].

The ordering of branches in vector V is as follows:

$$V = \begin{bmatrix} V_m \\ V_n \end{bmatrix} \tag{5.4}$$

where

V_n is the *voltage state vector* of order n. It is the voltage vector of tree branches.
V_m is the voltage vector of cotree branches.

Vector V_m can be found using voltage law Eq. (5.3) written as follows:

$$[1^m\quad L_n]\begin{bmatrix} V_m \\ V_n \end{bmatrix} = 1^m V_m + L_n V_n = 0 \tag{5.5}$$

Solving for V_m gives

$$V_m = -L_n\ V_n \tag{5.6}$$

Substituting L_n from Eq. (3.21), $L_n\ = -C_m^{\mathrm{T}}$

$$V_m = C_m^{\mathrm{T}}\ V_n \tag{5.7}$$

Finally, taking into account Eq. (5.4),

$$V = C^{\mathrm{T}}\ V_n \tag{5.8}$$

The voltage vector of tree branches V_n is the *voltage state vector.* Knowing vector V_n, the voltage vector of cotree branches can be found from Eqs. (5.6) and (5.7), called the *voltage state equations.* Equation (5.8) is the particular expression of voltage law in Eq. (5.3).

Physically it means that every element of voltage vector V is the product of transposed cut-set matrix and voltage state vector. Note the following difference between above equations: the voltage law Eq. (5.3) and voltage state Eq. (5.6) are expressed using tree loop-set matrix but the same kind voltages; voltage law Eq. (5.8) and voltage state Eq. (5.7) are expressed using tree cut-set matrix.

5.2 Linear Space Model of Network Voltages

In this section, the voltage vectors are modeled in the b-dimensions linear space $\bar{A}$. The set of all b-dimensional branches voltage vectors V satisfying Eq. (5.3) for any matrix L is a linear subspace $\overline{V}$ of the space $\bar{A}$, and is called the *voltage space* $\overline{V}$. From Eq. (5.3), it follows that the voltage subspace $\overline{V}$ is orthogonal to the cut-set subspace $\overline{C}$. It means that subspace $\overline{V}$ coincides with the subspace $\overline{C}$. Taking into account this coincidence, the subspace $\overline{V}$ is spanned by the basis of n cut-set vectors C [see Eq. (3.11)].

The linear space model of network voltages can be illustrated by a very simple example of 3-branch network in Fig. 5.1. The following topology of network is chosen. Branches 1 and 3 are a tree, and branch 2 is a cotree. The cut-set vectors defined by $m = 2$ tree branches are as follows: cut-set vector C_1 (generated by tree branch 1) and cut-set vector C_3 (generated by tree branch 3). Vectors C_1 and C_3 span the two-dimensional cut-set subspace $\overline{C}$ (plane). The loop-set vector defined by $n = 1$ cotree branch is loop-set vector L_2 (generated by cotree branch 2) which spans the one-dimensional loop-set subspace $\overline{L}$ (line).

Assuming the values of branch voltages, $v_1 = v_2 = 1$ and $v_3 = 2$, the voltage vector $V = [1 \quad 1 \quad 2]$ and spans the voltage subspace $\overline{V}$ (plane). The voltage subspace $\overline{V}$ is orthogonal (perpendicular) to subspace $\overline{L}$.

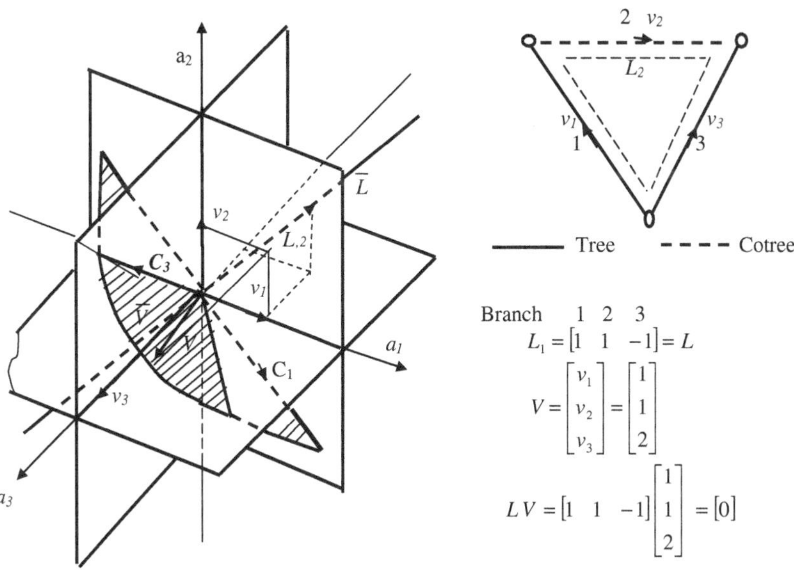

Fig. 5.1 Algebraic model of voltage space $\overline{V}$ for a simple example of 3-branch network

5.3 Topological Transformation of Voltage Space

So far, the voltages of network system are modeled in space $\bar{A}$. Taking into account that the cut-set vectors and loop-set vectors span a new b-dimensional linear space $\overline{T}$ (see Chap. 3, Sect. 3.5), the network voltage vector can also be expressed in a new basis, namely basis of topological space $\overline{T}$. It means that there exists linear transformation of vectors from space $\overline{T}$ to space $\bar{A}$. The transformation matrix is matrix T (see Sect. 3.5). Voltage vector V is defined in the old basis of space $\bar{A}$. As it was shown in Sect. 3.5, the topological matrix T is the transformation matrix of vectors between spaces $\bar{A}$ and $\overline{T}$. In order to find the voltage vector V (defined in space $\bar{A}$), transformed into space $\overline{T}$, let us write vector V as follows:

In space $\overline{T}$, let us write vector V as follows:

$$V = V_{m,o} + V_{o,n} = \begin{bmatrix} V_m \\ 0 \end{bmatrix} + \begin{bmatrix} 0 \\ V_n \end{bmatrix} = \begin{bmatrix} V_m \\ V_n \end{bmatrix} \tag{5.9}$$

where $V_{m,o}$ and $V_{o,n}$ are cotree and tree branch voltage vectors of order b (see Eq. (5.4))

It can be easily proved that the voltage vector $V_{o,n}$ after topological transformation from space $\overline{T}$ into space $\bar{A}$ is the voltage vector V. Starting from Eq. (5.8) —$V = C^{\mathrm{T}} V_n$ and substituting in this equation the tree voltage vector V_n of order n by the tree voltage vector $V_{o,n}$ of order b and substituting the cut-set matrix C^{T} by the matrix of topological transformation $T = \begin{bmatrix} L^{\mathrm{T}} & C^{\mathrm{T}} \end{bmatrix}$ we have:

$$V = \begin{bmatrix} L^{\mathrm{T}} & C^{\mathrm{T}} \end{bmatrix} V_{o,n} \tag{5.10}$$

Note that

$$\begin{bmatrix} L^{\mathrm{T}} & C^{\mathrm{T}} \end{bmatrix} V_{o,n} = C^T V_n : \tag{5.11}$$

It means that voltage vector $V_{o,n}$ is the vector of topological space $\overline{T}$ and if transformed using matrix T is the voltage vector V of space $\bar{A}$.

$$V = T\, V_{o,n} \tag{5.12}$$

All vectors $V_{o,n}$ form the subspace $\overline{V_{o,n}}$ in topological space $\overline{T}$ Both vectors V and $V_{o,n}$ are the same current vector in two spaces: $V_{o,n}$ in subspace $\overline{V}$ of space $\bar{A}$ and $V_{o,n}$ in subspace $\overline{V_{o,n}}$ of space $\overline{T}$. Note that vector $V_{o,n}$ is b-dimensional, and because of the orthogonality of the spaces $\overline{L}$ and $\overline{V}$, vector $V_{o,n}$ cannot involve any linear combinations of the columns of matrix C, so $V_{o,n}$ has only m nonzero components. Both the vectors $V_{o,n}$ and V_n determine the current state of network, and each of them is called the *voltage state vectors*. They differ by dimension only, $V_{o,n}$ is b-dimensional vector, and V_n is n-dimensional vector.

Illustrating the relation (5.12), in case of three-dimensional current space in Fig. 5.1, let us substitute into (5.12):

$$V = \begin{bmatrix} 1 \\ 1 \\ 2 \end{bmatrix}, \quad V_{o,n} = \begin{bmatrix} 0 \\ 1 \\ 2 \end{bmatrix}, \quad T = \begin{bmatrix} 1 & -1 & 1 \\ 1 & 1 & 0 \\ -1 & 0 & 1 \end{bmatrix} \tag{5.13}$$

Equation (5.12) is as follows:

$$\begin{array}{ccccc} & & \begin{matrix} 2 & 1 & 3 \end{matrix} & & \\ \begin{bmatrix} 1 \\ 1 \\ 2 \end{bmatrix} & = & \begin{bmatrix} 1 & -1 & 1 \\ 1 & 1 & 0 \\ -1 & 0 & 1 \end{bmatrix} & \begin{bmatrix} 0 \\ 1 \\ 2 \end{bmatrix} \\ V & = & T & V_{o,n} \end{array}$$

5.4 Voltage Relations Using Topological Matrix *T*

Using the topological transformation matrix *T*, various interrelations between the *b*-dimensional current vectors can be derived. The starting point is Eqs. (4.10) and (4.12).

Using the topological transformation matrix *T*, various interrelations between the *b*-dimensional voltage vectors can be derived. The starting point is Eqs. (5.10) and (5.12).

Matrix *T* is non-singular, so the voltage state vector $V_{o,n}$ can be found using Eq. (4.12)

$$V_{o,n} = T^{-1}\,V \tag{5.14}$$

Interpreting physically the above equation, it means that the set of network voltages, which are dependent on each other (must fulfill the voltage law), can be substituted and considered using the subset of these network voltages, which are independent of each other.

The relation between $V_{o,n}$ and $V_{m,o}$ comes as a result of substituting *V* of Eq. (5.9) into (5.12).

$$V_{m,o} = (T - 1)\,V_{o,n} \tag{5.15}$$

This is just *b*-dimensional version of relation (5.7). Matrix $(T - 1)$ is usually singular, so if the cotree voltages are known, one cannot usually find the tree voltages. The substitution of Eq. (5.10) into (5.14) gives the relation between $V_{m,o}$ and *V*:

$$V_{m,o} = (1 - T^{-1})\,V \tag{5.16}$$

An interesting voltage vector can be derived using the transposed topological matrix T^{T}.

$$T^{\mathrm{T}}\,V = V_n^C \tag{5.17}$$

where V_n^C can be called *the cut-set voltage vector.* In order to show that vector V_n^C has physical meaning, note that

$$V_n^C = T^{\mathrm{T}} \quad V = \begin{bmatrix} L \\ C \end{bmatrix} \quad V = C\,I \quad \text{since} \quad L\,V = 0 \tag{5.18}$$

Hence, each entry of V_n^C is the algebraic sum of the branch voltages of the cut-set, defined by a tree branch. Physically, it means that having n numbers, each of them being the algebraic sum of branch voltages of one of n tree cut-sets, one can find all branch voltages in the network. Inverting Eq. (5.16), we have

$$V = (T^{\mathrm{T}})^{-1}\,V_n^C \tag{5.19}$$

Substituting Eq. (5.10) into (5.17) gives the relation between V_n^C and $V_n^{'}$

$$V_n^C = (T^{\mathrm{T}}T)^{-1}\,V_{o,n} \tag{5.20}$$

From Eqs. (5.17) and (5.18), it follows that the voltage vector V_n^C uniquely defines the current state of network. Substituting Eq. (5.17) into (5.14) gives the relation between $V_{m,o}$ and V_n^C

$$V_{m,o} = (1 - T^{-1})\,(T^{\mathrm{T}})^{-1} V_n^C \tag{5.21}$$

The above-derived formulas do not exhaust all possible. Reducing the relations to the order m and n enables derivation of various practically useful equations.

Chapter 6
Algebraic Model of Current–Voltage Vectors

6.1 Orthogonality of Current and Voltage Vectors

The algebraic models of currents and voltages have been derived in the previous chapters. In this chapter, the important physical notion—the combined current–voltage vector—is defined and described. Analogically as the current and the voltage models, the current–voltage model is derived in terms of linear algebra. In order to show the background of current–voltage quantity, let us start from the power law. The *branch power* s_j in the *j*th branch of network can be defined as follows:

$$s_j = i_j^* v_j \quad \text{or} \quad s_j = v_j^* i_j \tag{6.1}$$

where i_j and v_j are the branch current and branch voltage in the *j*th branch, and i_j^* and v_j^* are the complex conjugates of i_j and v_j. The *power law of network* states that the sum of all branch powers at any instant is equal to zero. In other words, for any feasible current and voltage vectors I and V

$$I^* V = \sum_{j-1}^{j=b} s_j = 0 \tag{6.2}$$

where I^* is the conjugate transpose of the branch current I.

This means that the current subspace $\overline{I}$ and the voltage subspace $\overline{V}$ are orthogonal if the same branch ordering is used in both spaces.

The Eq. (6.2) is sometimes taken as axiomatic, but it can be proved that it is a consequence of orthogonality of the cut-set and loop-set matrices as follows: Suppose I and V are given in the form of Eqs. (4.8) and (5.8) as follows:

$$I = L^T I_m \quad \text{and} \quad V = C^T V_n$$

A. Kłos, *Mathematical Models of Electrical Network Systems*,
Lecture Notes in Electrical Engineering 412,
DOI 10.1007/978-3-319-52178-7_6

Substituting these equations to (6.2) we have:

$$I^{*}V = (L^{T}I_{m})^{*}C^{T}V_{n} = I_{m}^{*}(L^{T})^{*}C^{T}V_{n} \tag{6.3}$$

Since L is a real matrix, $(L^{T})^{*} = (L^{T})^{T} = L$, then

$$I^{*}\,V \ = I_{m}^{*}\,L\,C^{T}\,V_{n} \tag{6.4}$$

Taking into account the orthogonality of the cut-set and loop-set matrices $L\,C^{T} = 0$ [see Eq. (3.14)], the above equation is just the power law

$$I^{*}\,V \ = 0 \tag{6.5}$$

So it was proved that the current subspace $\overline{I}$ and the voltage subspace $\overline{V}$ are orthogonal to each other.

6.2 Current–Voltage Vector

In this section, summarizing the results of current and voltage algebraic models given in Chaps. 4 and 5, the current–voltage vector is defined.

Taking into account that the current state vectors $I_{m,o}$ and voltage state vectors $V_{o,n}$ [see Eqs. (4.9) and (5.9)] are vectors in the topological space $\overline{T}$ and are orthogonal to each other, they can be added together in the current–voltage state vector.

The *current–voltage state vector* is defined as a sum of vectors $I_{m,o}$ and $V_{o,n}$, denoted as $K_{i,v}$

$$K_{i,v} = I_{m,o} + V_{on} = \begin{bmatrix} I_m \\ 0 \end{bmatrix} + \begin{bmatrix} 0 \\ V_n \end{bmatrix} = \begin{bmatrix} I_m \\ V_n \end{bmatrix} \tag{6.6}$$

where $K_{i,v}$ is a vector of space $\overline{T}$ of order b.

The current–voltage space $\overline{K}_{i,v}$ is the set of all current–voltage vectors $K_{i,v}$.

Vector $K_{i,v}$ may be unexpected from the physical point of view but is natural from the mathematical viewpoint. It is not widely known but it is linked to a familiar vector, which includes the current and voltage state vectors and which is used in classical network analysis.

In order to justify the physical meaning of the summation of currents and voltages, let us use the topological transformations of current vectors and of voltage vectors described in Chaps. 4 and 5. Using topological transformation matrix T, the current and voltage state vectors $I_{m,o}$ and $V_{o,n}$ are transformed from topological space $\overline{T}$ into space $\overline{A}$ [see Eqs. (4.14) and (5.14)] and are the current and voltage vectors I and V. Substituting Eqs. (4.14) and (5.14) into Eq. (6.6) we have:

$$K_{i,v} = T^{-1}I + T^{-1}V = T^{-1}(I + V) \tag{6.7}$$

where $(I + V)$ is a new, non-conventional, and interesting vector (as it will be shown in what follows). Denoting this vector K:

$$K = I + V, \tag{6.8}$$

the relation between the current state vector $K_{i,v}$ and network current vector I is as follows:

$$K_{i,v} = T^{-1}K \tag{6.9}$$

where

The current–voltage vector K is defined as the sum of the current vector I and the voltage vector V in the network.

Relation (6.9) means that in terms of linear algebra, $K_{i,v}$ is the topological transformations of current–voltage vector K.

Every element of vector K is the sum of branch current i_j and branch voltage v_j

$$k_j = i_j + v_j. \tag{6.10}$$

The current–voltage subspace $\overline{K}$ of space $\overline{A}$ is the set of all current–voltage vectors K. Subspace $\overline{K}$ includes the spaces $\overline{I}$ and $\overline{V}$, and is the directed sum of subspaces $\overline{I}$ and $\overline{V}$. The vectors I and V may be recovered as the projections of $\overline{K}$ into $\overline{I}$ and $\overline{V}$.

The current–voltage vector K as a *sum* of vectors I and V is unexpected from the physical point of view but is natural from the mathematical viewpoint.

6.3 Linear Space Model of Current–Voltage Vectors

The current–voltage state space of a simple example of 3-branch network is illustrated in Fig. 6.1. In Fig. 6.1a, the spaces $\overline{I}$, $\overline{V}$, and $\overline{K}$ are subspaces of space $\overline{A}$. Vectors I, V and K are numerically defined in orthonormal basis (a_1, a_2, a_3) of space $\overline{A}$.

In Fig. 6.1b, the spaces $\overline{I}_{m,o}$, $\overline{V}_{o,n}$, and $\overline{K}_{i,v}$ are subspaces of space $\overline{T}$. Vectors $I_{m,o}$, V_{on}, and K are numerically defined in cut-set and loop-set basis (C_1, C_3, L_2) of space $\overline{T}$.

Let us illustrate the numerical values of current–voltage vectors using the branch currents and branch voltages of simple examples of 3-branch network in Chaps. 4 and 5.

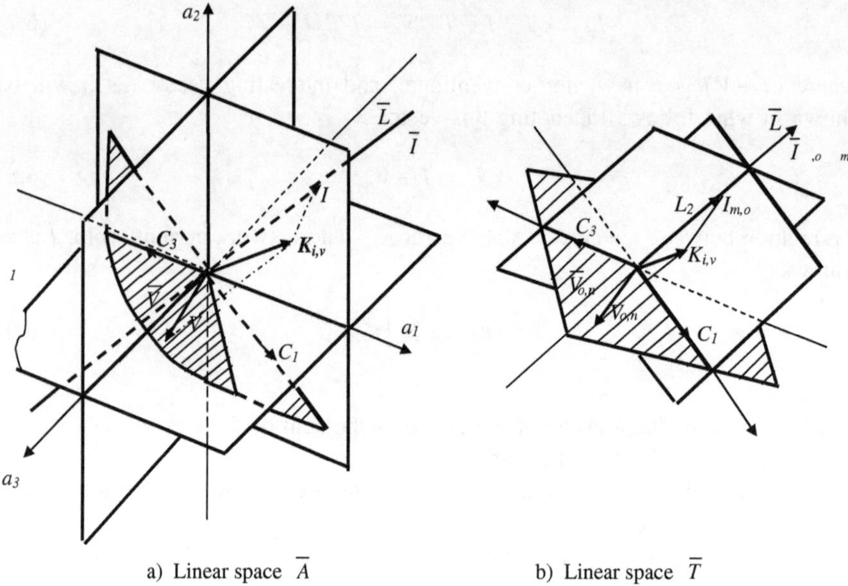

Fig. 6.1 The current–voltage vectors in linear spaces $\overline{A}$ and $\overline{T}$. **a** Linear space $\overline{A}$. **b** Linear space $\overline{T}$

$$I = \begin{bmatrix} 1 \\ 1 \\ -1 \end{bmatrix}, I_{m,o} = \begin{bmatrix} 1 \\ 0 \\ 0 \end{bmatrix}, V = \begin{bmatrix} 1 \\ 1 \\ 2 \end{bmatrix}, V_{o,n} = \begin{bmatrix} 0 \\ 1 \\ 2 \end{bmatrix}, T = \begin{bmatrix} 1 & -1 & 1 \\ 1 & 1 & 0 \\ -1 & 0 & 1 \end{bmatrix}, T = \begin{bmatrix} 1/3 & 1/3 & -1/3 \\ -1/3 & 2/3 & 1/3 \\ 1/3 & 1/3 & 2/3 \end{bmatrix},$$

$$K_{i,v} = I_{m,o} + V_{o,n} = \begin{bmatrix} 1 \\ 0 \\ 0 \end{bmatrix} + \begin{bmatrix} 0 \\ 1 \\ 2 \end{bmatrix} = \begin{bmatrix} 1 \\ 1 \\ 2 \end{bmatrix}, \quad K = I + V = \begin{bmatrix} 1 \\ 1 \\ -1 \end{bmatrix} + \begin{bmatrix} 1 \\ 1 \\ 2 \end{bmatrix} = \begin{bmatrix} 2 \\ 2 \\ 1 \end{bmatrix}$$

$$K = T\,K_{i,v} = \begin{bmatrix} 1 & -1 & 1 \\ 1 & 1 & 0 \\ -1 & 0 & 1 \end{bmatrix} \begin{bmatrix} 1 \\ 1 \\ 2 \end{bmatrix} = \begin{bmatrix} 2 \\ 2 \\ 1 \end{bmatrix}, \quad K_{i,v} = T^{-1}K \begin{bmatrix} 1/3 & 1/3 & -1/3 \\ -1/3 & 2/3 & 1/3 \\ 1/3 & 1/3 & 2/3 \end{bmatrix} \begin{bmatrix} 2 \\ 2 \\ 1 \end{bmatrix} = \begin{bmatrix} 1 \\ 1 \\ 2 \end{bmatrix}$$

6.4 Current–Voltage Vectors' Relations Using Matrix T

The equations for the recovery of current and voltage state vectors from either K or K_{iv} are using the order $m\ x\ b$ and $n\ x\ b$ matrices $[\,1^m \quad 0\,]$ and $[\,0 \quad 1^n\,]$ to pick out the first m and the last n components of matrix K_{iv}

$$I_m = [\,1^m \quad 0\,]K_{iv} \quad V_n = [0 \quad 1^n]K_{i\,v} \tag{6.11}$$

Using this and equations ( ), ( ) and assuming all matrices have matching row and column orderings

$$I = L^T[1^m \; 0]K_{i,v} = L^T[1^m \; 0]T^{-1}K \tag{6.12}$$

$$V = C^T[0 \; 1^n]K_{i,v} = C^T[0 \; 1^n]T^{-1}K \tag{6.13}$$

Let us define a new vector opposite to the current–voltage vector denoted as $K_{v\,i}$, which is the sum of cotree voltages and tree currents vectors.

The *voltage–current vector* $K_{v\,i}$ is the sum of cotree voltages and tree currents vectors:

$$K_{v\,i} = \begin{bmatrix} V_m \\ 0 \end{bmatrix} + \begin{bmatrix} 0 \\ I_n \end{bmatrix} = \begin{bmatrix} V_m \\ I_n \end{bmatrix} \tag{6.14}$$

Using Eqs. (4.7) and (5.7), the relation between $K_{v\,i}$ and $K_{i,v}$ is as follows:

$$K_{vi} = (T - 1)K_{iv} \tag{6.15}$$

Analogically using Eqs. (4.16) and (5.16), the relation between $K_{v\,i}$ and K is as follows:

$$K_{vi} = (1 - T^{-1})K \tag{6.16}$$

Using matrix T^T and summing the vectors I_m^L and V_n^C [see Eqs. (4.17) and (5.17)], a new current–voltage vector $K^{L,C}$ is as follows:

$$K^{L,C} = T^T K = I_m^L + V_n^C = I_m^L + V_n^C \tag{6.17}$$

where each of n entries of $K^{L,C}$ is the algebraic sum of the branch currents along the loop-set, and each of m entries of $K^{L,C}$ is the algebraic sum of the branch voltages belonging to the cut-set. Inverting Eq. (6.17), we have:

$$K = (T^T)^{-1}K^{L,C} \tag{6.18}$$

Note that vector $K^{L,C}$ is the topological transformation of vector K.

6.5 Comment

Up to this chapter, the modeling process has produced formulae using more general approach to the mathematical modeling of electrical networks. In this chapter, as happens frequently, the model has produced unexpected results concerning the current–voltage vector K, which do not arise naturally from physics. The current–voltage state vector K appears as natural in the linear space model formulated here; yet physically, this vector is new and unexpected in network theory. However, it means that mathematical summation of physical quantities may have physical meaning. The addition of currents and voltages is not an obvious process to carry out, especially as the units of measurements do not matter as long as they are consistent for all currents and for all voltages. For example, currents might be in amperes and voltages in volts. The vector $K_{i,v}$, on the other hand, is often used in physics and should present no difficulties to circuit theorists because it uniquely defines the current and voltage state of network. However, the vector K does the same thing and so has an equal right to be called a state vector. In terms of linear algebra, $K_{i,v}$ and K are just two different column matrices representing the same current–voltage vector with respect to two different bases as was remarked previously in Chaps. 4 and 5.

Chapter 7
Kirchhoff's Laws Using Matrix T

One of the main results of the algebraic models of current and voltage states, derived in previous chapters, is the introduction of two b-dimensional linear spaces (b—number of network branches). It makes possible derivation of various current, voltage, current–voltage relations and formulae in two different and isomorphic linear spaces: space $\overline{A}$ and space $\overline{T}$. This was done in Chaps. 4, 5, and 6 using classical formulation of the Kirchhoff's laws and topological matrix T. However, it appears that modeling process is leading to the non-conventional formulations of Kirchhoff's laws. In this chapter, three of such formulations are presented.

7.1 Basic b-Dimensional Formulation

In Chaps. 4 and 5, the current and voltage vectors of network are expressed in two linear spaces: the current and voltage vectors I and V in the space $\overline{A}$, and the current and voltage state vectors $I_{m,o}$ and $V_{o,n}$ in the topological basis of space $\overline{T}$. The relations between these expressions are as follows (see Eqs. 4.14 and 5.14):

$$I_{m,o} = T^{-1}I \tag{7.1}$$

$$V_{o,n} = T^{-1}V \tag{7.2}$$

where

$$I_{m,o} = \begin{bmatrix} I_m \\ 0 \end{bmatrix} \quad \text{and} \quad I = \begin{bmatrix} I_m \\ I_n \end{bmatrix}$$

$$V_{o,n} = \begin{bmatrix} 0 \\ V_n \end{bmatrix} \quad \text{and} \quad V = \begin{bmatrix} V_m \\ V_n \end{bmatrix}.$$

T is the topological transformation matrix.

A. Kłos, *Mathematical Models of Electrical Network Systems*,
Lecture Notes in Electrical Engineering 412,
DOI 10.1007/978-3-319-52178-7_7

Note that vector I_m is on both sides of Eq. (7.1) and vector V_n is on both sides of Eq. (7.2). Let us transpose in Eq. (7.1) a vector I_m from left to right hand side, what means adding number -1 to the m upper diagonal elements of matrix T^{-1}, and analogically; transpose in Eq. (7.2) a vector V_n from left- to right-hand side, what means adding –1 to the n lower diagonal elements of matrix T^{-1}. As a result of such transpositions, matrix T^{-1} in Eqs. (7.1 and 7.2) changes as follows:

In current equation (7.1), the new matrix denoted $(T^{-1})_I$

$$(T^{-1})_I = \begin{bmatrix} T_1^{-1} - 1^m & T_2^{-1} \\ T_3^{-1} & T_4^{-1} \end{bmatrix} \quad (7.3)$$

In voltage equation (7.2), the new matrix denoted $(T^{-1})_V$

$$(T^{-1})_V = \begin{bmatrix} T_1^{-1} & T_2^{-1} \\ T_3^{-1} & T_4^{-1} - 1^n \end{bmatrix} \quad (7.4)$$

where T_j^{-1} are submatrices of matrix T^{-1} and 1^m, 1^n are unit matrices of order m and n.

The current and voltage laws (7.1 and 7.2) take the homogenous form as follows:

$$(T^{-1})_I \, I = 0 \quad (7.5)$$

$$(T^{-1})_V \, V = 0 \quad (7.6)$$

Let us now change matrices $(T^{-1})_I$ and $(T^{-1})_V$ as follows: In both matrices, each entry of each row is divided by a diagonal element of this row. Such division does not change the homogenous equations (the spaces of solutions remain unchanged) because all diagonal entries of matrix T^{-1} are not equal to zero. As a result, the new matrices are denoted as follows: A_I called *current law matrix* and A_V called *voltage law matrix*. The current and voltage laws (7.1 and 7.2) are as follows:

$$A_I \, I = 0 \quad (7.7)$$

$$A_V \, V = 0 \quad (7.8)$$

Equations (7.7 and 7.8) are a general formulation of the current and voltage laws and can be called *the b-dimensional formulation of Kirchhoff's laws.* They are analogy to the classical formulations $C\,I = 0$ but A_I and A_V are not incidence matrices. The matrices A_I and A_V are of order $b \times b$ and of rank m and n accordingly. The values of elements of A_I and A_V depend on the structure of network graph and the diagonal elements are equal to 1. Generally for a given network, the matrices A_I and A_V are not unique and depend on the topology of network.

However, independent of tree–cotree topology, certain elements in all matrices A_I and A_V (the same elements in both matrices) are equal to zero.

7.2 Pseudo-Unit Formulation

Another formulation of Kirchhoff's laws can be derived starting from the b-dimensional formulation (Eqs. 7.7 and 7.8). Let us change matrix A_I as follows: All diagonal elements of matrix A_I, which are equal to 1, are transferred from the left to the right side of Eq. (7.7). So the diagonal elements of A_I are equal to 0. Such matrix, with minus sign, is denoted B_I.

$$B_I = -A_I + 1^b \tag{7.9}$$

Doing the same operation with matrix A_V, we have the following:

$$B_V = -A_V + 1^b \tag{7.10}$$

After this operations, matrices A_I and A_V are as follows:

$$A_I = -B_I + 1^b \tag{7.11}$$

$$A_V = -B_V + 1^b \tag{7.12}$$

Substituting Eqs. (7.11 and 7.12) into (7.7 and 7.8), we have final equations as follows:

$$B_I\, I = I \tag{7.13}$$

$$B_V\, V = V \tag{7.14}$$

Matrices B_I and B_V are a sort of pseudo-unit matrices. So the Eqs. (7.13 and 7.14) can be called *the pseudo-unit formulation of Kirchhoff's laws*.

Note that the diagonal elements of the matrices B_I and B_V are equal to zero. So each row in Eq. (7.13) is a function relating one branch current to all remaining branch currents and each row in Eq. (7.14) is a function relating one branch voltage with all remaining branch voltages. Note that, while in the classical network analysis, such relations are done using simple (0, 1, −1) incidence matrices C and L, and then in pseudo-unit formulation, one current and one voltage are the sums of all other currents and the sums of other voltages, with coefficients equal to the elements of matrices B_I and B_V. Generally, the matrices B_I and B_V are not unique. The investigation done show that its uniqueness depends on the structure of network graph. However, for some networks, e.g., for full graph networks, there is only one matrix B_I and one B_V. For majority of other network graphs, there is a number (it depends on network graph) of different matrices B_I and B_V, but in all of them, some rows and some entries are identical.

7.3 Current–Voltage Formulation

As it is shown in Chap. 6, the current–voltage state vector $K_{i,v} = I_{m,o} + V_{o,n}$ is the topological transformations of current–voltage vector $K = I + V$.

$$K_{m,n} = T^{-1}\,K \tag{7.15}$$

It means that Eq. (7.15) contains both current and voltage Kirchhoff's laws. Practically, it means that knowing the sum of current and voltage values in each network branch, and knowing in what units the summation is done, one can find, using matrix T, all branch currents and branch voltages in the network. The Eq. (7.15) can be called *the combined current–voltage formulation of Kirchhoff's law*.

Note that the summation of current and voltage may have applications in measuring and information technology, e.g., measuring together (instead of separate) a sum of branch current and voltage may change the measurement systems in power system networks.

Chapter 8
Current–Voltage Functional Relations

8.1 Branch Parameters

Up to this point, the branch currents and branch voltages of a network system have been assumed to be unrelated, except that the current space $\overline{I}$ and the voltage space $\overline{V}$ are orthogonal and that individual current and voltage vectors obey the power law $I^{*}V = 0$ (see Chap. 6). This means that current vector I and voltage vector V are independent on each other, as long current I remains in current subspace $\overline{I}$ and voltage V in voltage subspace $\overline{V}$. In physical networks, this is not the case and the branch currents and branch voltages are related each other. These relations, generally called *the branch current–voltage functional relations* (in what follows called *current–voltage functions*) are considered in this chapter. In physical networks, they take on many different forms of which Ohm's law is a special type, which often occurs in electrical networks. The most general form of the current–voltage functional relations, in the b-branch network, would be represented by a set of mathematical functions of $2b$ variables

$$f(I,V) = f(i_1, i_2, \ldots i_b,\ v_1, v_2, \ldots v_b) \tag{8.1}$$

where function $f(I,V)$ may be very different.

In electrical network systems, the $f(I,V)$ functions depend on the physical nature of branches.

In this work, the considered electrical network systems are restricted to being time constant and linear or nonlinear of the power-type relation $i_j v_j = p_j$. In this chapter, the current–voltage functions of network system branches are described using Ohm's law. In all the network systems, the network branches initiate (generate) the current state and voltage state of network and determine the relations between current values and voltage values. It means that the network system must include branches which are: current or/and voltage sources and current–voltage functions Mathematically, to each branch the linear function is associated, called

A. Kłos, *Mathematical Models of Electrical Network Systems*,
Lecture Notes in Electrical Engineering 412,
DOI 10.1007/978-3-319-52178-7_8

the *branch equation* or *branch parameter*. There are three kinds of branch equations (branch parameters):

- First kind is the branch equations, which are branch current–voltage functions. They transform branch current into branch voltage or vice versa, using constant coefficients and are called *branch admittance*, denoted y_j or *branch impedance*, denoted z_j.
- Second kind is the branch equations, which generate the currents or voltages in a network. They are the constant current or constant voltage sources and are called *the ideal current sources* and *the ideal voltage sources*.
- Third kind is the branch equations, which are both branch current–voltage functions and current source and/or voltage source.

In the classical network analysis, if the inductive and capacitive couplings between branches (self- and mutual couplings) are to be taken into account, then the branch equation j of third kind is written, traditionally either in admittance or impedance notation, as follows:

$$i_j = \sum_{k=1}^{k=b} (y_{j,k}\, v_k) + e \tag{8.2a}$$

or

$$v_j = \sum_{k=1}^{k=b} (z_{j,k}\, i_k) + e \tag{8.2b}$$

where

i_j, v_j current of branch j and voltage of branch j.
$y_{j,k}$, $z_{j,k}$ self- and mutual branch admittances and branch impedances, constant branch parameters.
e current source or/and voltage source, constant branch parameters.

Each of the above two notations of branch equations can be used as the full mathematical model of the branch equations of b-branch network; however, not both notations can be used together for a whole or a part of network, because $y_{j,k}$ and $z_{j,k}$ are the different notations only of the very same parameters of a branch. However, both notations can be used if all branches are partitioned into two parts.

The model of branch equations in this chapter refers to the algebraic model of network system presented in this book. If the network is topologically partitioned into m cotree branches and n tree branches, then, in what follows, we assume that the branch equations of a cotree branches are admittances (Eq. 8.2a) and the branch equations of a tree branches are impedances (Eq. 8.2b), ($m + n = b$). Note, however, that in such notation, we may have the current sources and the voltage sources in both equations. From algebraic viewpoint, the sources should be dislocated. The branch equations of cotree branches are currents, so they should include the current sources only and the branch equations of tree branches are voltages, so they should include the voltage sources only. It means that the branches with current sources in tree branches should be dislocated to the cotree and changed from impedance type into admittance type, and branches with voltage sources in cotree branches should be dislocated to the tree branches and changed from admittance type to impedance type. After such dislocations, the branch equations are as follows:

$$i_j = \sum_{k=1}^{k=m} (y_{j,k} v_k) + e^i \tag{8.3}$$

and

$$v_j = \sum_{k=1}^{k=n} (z_{j,k} i_k) + e^v \tag{8.4}$$

where

e^i current sources in cotree branch equation.
e^v voltage sources in tree branch equations.

Note that the numbers of current sources and voltage sources are restricted and imposed by current and voltage laws. The consistency condition of unique solution of a network problem is fulfilled if there are no more than m current sources and no more than n voltage sources in the network.

From the algebraic point of view, it may be of advantage to separate each current source and each voltage source in branch equations (8.3 and 8.4). Physically, this is equivalent to replace the branch equation with current source by its equivalent of two branch equations: first—active part of source in Eq. (8.3), now called *the ideal current source*; second—passive part of source in Eq. (8.3). Analogically replacing the branch equation with voltage source by its equivalent of two branch equations is the voltage source separation: first—active part of source in Eq. (8.4), now called *the ideal voltage source*; second—passive part of source in Eq. (8.4). It means additional branches in the network system. After separation, the branch equations are as follows:

$$i_j = e^i_j \qquad (j = 1, 2, 3, \ldots f') \tag{8.5}$$

$$i_j = \sum_{j=1}^{m} (y_{jk} \; v_k) \tag{8.6}$$

and

$$v_j = e^v_j \qquad (j = 1, 2, 3, \ldots f'') \tag{8.7}$$

$$v_j = \sum_{j=1}^{n} (z_{j,k} \, i_k) \tag{8.8}$$

where

f' is a number of ideal current sources.
f'' is a number of ideal voltage sources.

Branches, which are ideal sources, are a point-like branches, which do not have self- and mutual admittances or impedances. Generally in the network, the current branches (8.5 and 8.6) may be connected in series and the voltage branches (8.7 and 8.8) are often connected in parallel. Equations (8.5–8.8) may serve to the formulation of some special branch equations in the practical network solution problems. For example, the power system generator can be modeled as follows: ideal current source and ideal voltage source are connected in parallel and current–voltage functions are connected in series. The branch equation of some branches is often a nonlinear, complicated function; however, after linearization, the branch equation can be modeled using three types of equations: current source, voltage source, and current–voltage equations. In what follows, only the linear (Ohm's law) branch equations are used.

8.2 Ohm's Law of Network System—Algebraic Model

Each real network system must include at least one current or voltage source. Without any source, the network is empty ($I = 0$, $V = 0$) and the classical network Ohm's law in such network does not exist. The classical Ohms law can be used only to the parts of the network system.

In this section, the algebraic model of the Ohm's law in network system is derived, which takes into account the current and/or voltage sources.

Consider the network system which is a set of b interconnected branches. Suppose that any tree/cotree topological structure of network is chosen. Suppose that the branch equations are of both types (passive and active) and the Ohm's law is in the form of Eqs. (8.5–8.8). The network branches (branch equations) are ordered according to the order of m cotree and n tree branches. The ideal current sources are branch equations of cotree branches, so they are parts of current state vector, and the ideal voltage sources are branch equations of tree branches, so they are parts of voltage state vector V_n. The current state vector I_m which takes into account ideal current sources, now called as *the network-current state vector* and denoted as $I_{m,F}$, is as follows:

$$I_{m,F} = \begin{bmatrix} I_{m,I} \\ I_{m,y} \end{bmatrix} \tag{8.9}$$

where

$I_{m,I} = E_I$ is subvector of the network-current state vector of order f'' (f' number of sources); its elements are ideal current sources.

$I_{m,y}$ is the subvector of network-current state vector, which elements are cotree branch admittances, of order $m - f'$.

Analogically the voltage state vector V_n which takes into account ideal voltage sources, now called as the *network-voltage state vector* and denoted as $V_{n,F}$, is as follows:

$$V_{n,F} = \begin{bmatrix} V_{n,V} \\ V_{n,z} \end{bmatrix} \tag{8.10}$$

where

$V_{n,V} = E_V$ is subvector of the network-voltage state vector of order f'' (f'' number of sources), its elements are ideal voltage sources.

$V_{n,z}$ is the subvector of network-voltage state vector, which elements are tree branch impedances, of order $n - f''$.

The Ohm's law expressed in the form of Eqs. (8.5 and 8.8) can be written in the matrix form as follows:

$$I_{m,F} = \begin{bmatrix} I_{m,I} \\ I_{m,y} \end{bmatrix} = \begin{bmatrix} E_I \\ Y_{m,f'}\ V \end{bmatrix} \tag{8.11}$$

where

$Y_{m,f'}$ matrix of cotree branch admittances of order $m{-}f' \times b$.

$$V_{n,F} = \begin{bmatrix} V_{n,V} \\ V_{n,z} \end{bmatrix} = \begin{bmatrix} E_V \\ Z_{n,f''} I \end{bmatrix} \tag{8.12}$$

where

$Z_{n,f''}$ matrix of tree branch impedances of order $n{-}f'' \times b$

The Kirchhoff's current and voltage laws can be expressed in the form of Eqs. (4.8 and 5.8), which are derived in Chaps. 4 and 5.

$$I = L^T\, I_m \tag{8.13}$$

$$V = C^T\, V_n \tag{8.14}$$

Substituting Eq. (8.14) to (8.11) and Eq. (8.13) to (8.12), we have

$$I_{m,F} = \begin{bmatrix} I_{m,I} \\ I_{m,y} \end{bmatrix} = \begin{bmatrix} E_I \\ Y_{m,f'}\ C^T V_n \end{bmatrix} \tag{8.15}$$

$$V_{n,F} = \begin{bmatrix} V_{n,V} \\ V_{n,z} \end{bmatrix} = \begin{bmatrix} E_V \\ Z_{n,f''} L^T I_m \end{bmatrix} \tag{8.16}$$

where

$I_{m,F}$ *The current state vector of network system*, of order *m*.
$V_{n,F}$ *The voltage state vector of network system*, of order *n*.

Equations (8.15 and 8.16) are the algebraic model of the Ohm's law of network system.

Note that both equations satisfy the Kirchhoff's and the Ohm's laws and are the mathematical relations between the current and voltage vectors: $I_{m,I}, I_{m,y}, V_{n,V}, V_{n,z}$ and branch parameters: $E_I, E_V, Z_{n,f''}, Y_{m,f'}$ in network system. If all branch equations are the current–voltage functions only ($f' = 0$ and $f'' = 0$), then Eqs. (8.15 and 8.16) reduce to equations:

$$I_m = Y_m\, C^T\, V_n \tag{8.17}$$

$$V_n = Z_n\, L^T\, I_m \tag{8.18}$$

which are the classical Ohm's law in the form of relations: current state vector as a function of voltage state vector or vice versa (matrices Y_m, Z_n are restricted: Y_m is of order $m \times b$ and Z_n is of order $n \times b$).

Note that in Eqs. (8.15 and 8.16), the network-current state vector $I_{m,F}$ is a function of voltage state vector V_n and network-voltage state vector $V_{n,F}$ is a function of current state I_m. It means that if the network topology and branch admittances Y_m are known, then choosing any voltage state vector V_n, the network-current state vector $I_{m,F}$ can be found. Analogically, if network topology and branch impedances Z_n are known, then choosing any current state vector I_m, the network-voltage state vector $V_{n,F}$ can be found.

8.3 Linear Space Model of Current–Voltage State Vectors in Network System

In Chap. 6, the algebraic model of the current–voltage state vector $K_{i,v}$ and the linear state space $\overline{K_{i,v}}$ of all vectors $K_{i,v}$ are defined. So far vector $K_{i,v}$ and space $\overline{K_{i,v}}$ are derived using Kirchhoff's laws only and are illustrated in the form of a linear space on a simple example of network in Fig. 6.1 (see Sect. 6.3). In this section, the current–voltage vector is defined, which satisfy Kirchhoff's and Ohm's laws. Let us start from definition of vector $K_{i,v}$ given in Chap. 6.

$$K_{I,V} = I_{m,o} + V_{o,n} = \begin{bmatrix} I_m \\ 0 \end{bmatrix} + \begin{bmatrix} 0 \\ V_n \end{bmatrix} = \begin{bmatrix} I_m \\ V_n \end{bmatrix} \tag{8.19}$$

Substituting in Eq. (8.19) vector I_m by its image $I_{m,F}$ from Eq. (8.15) and vector V_n by its image $V_{n,F}$ from Eq. (8.16), we have

$$K_{I,V,F} = \begin{bmatrix} I_{m,F} \\ V_{n,F} \end{bmatrix} = \begin{bmatrix} I_{m,I} \\ I_{m,y} \\ 0 \\ 0 \end{bmatrix} + \begin{bmatrix} 0 \\ 0 \\ V_{n,V} \\ V_{n,z} \end{bmatrix} = \begin{bmatrix} E_I \\ Y_{m,F}C^T V_n \\ E_V \\ Z_{n,F}L^T I_m \end{bmatrix} \tag{8.20}$$

where

$K_{I,V,F}$ is *the current–voltage state vector of network system.*

All vectors $K_{I,V,F}$ form the *current–voltage space of network system* $\overline{K_{I,V,F}}$. All vectors $I_{m,F}$ and all vectors $V_{o,F}$ form, accordingly the *current space of network system* $\overline{I_{m,F}}$ and the *voltage space of network system* $\overline{V_{n,F}}$. Note that Eq. (8.20) represents both the Kirchhoff's and the Ohm's network system laws.

In what follows, the *current–voltage space of network system* $\overline{K_{I,V,F}}$ is illustrated on a simple example of 3-branch network. It is shown that ideal current and voltage sources are reducing the dimensions of spaces, $\overline{I_{m,F}}$, $\overline{V_{o,F}}$ and space $\overline{K_{I,V,F}}$.

1. If in the network the ideal current sources do exist and there are no ideal voltage sources, then in Eq. (8.15) matrix $Y_{m,F}$ is restricted (is of order $m - f' \times b$) and network-current state vector $I_{m,F}$ includes f' constant ideal current sources (includes subvector $I_{m,I}$). It means that the network-current–voltage state vectors $K_{I,v,F}$ include f' constant elements (ideal current sources). So the space $\overline{K_{I,V,F}}$ is of dimension $b - f'$. It is illustrated in Fig. 8.1. Branch *2* (one cotree branch) is ideal current source, so vector $I_{m,I}$ is equal constant value $= E_{I2}$, and space $\overline{I_{m,F}}$ is of dimension zero. Branches *1* and *3* (two branches of a tree) are impedance branches, so $V_{n,F} = \begin{bmatrix} V_{n,F1} \\ V_{n,F3} \end{bmatrix}$ and space $\overline{V_{o,F}}$ is of dimension 2 [plane (a_1, a_3)]. It means that the dimension of network space $\overline{K_{I,V,F}}$ is equal $b - f' = 2$ and is shown as the two-dimensional lattice plane in Fig. 8.1.
2. Analogically, if the voltage sources do exist and there are no ideal current sources, then in Eq. (8.15) matrix $Z_{n,F}$ is restricted (is of order $m - f'' \times b$) and network-voltage state vector $V_{n,F}$ includes f'' constant ideal voltage sources (includes subvector $V_{n,V}$). It means that the network-current–voltage state vectors $K_{I,v,F}$ include f'' constant elements. So the space $\overline{K_{I,V,F}}$ is of dimension $b - f$ '. It is illustrated in Fig. 8.2. Branch number *1* (one of two tree branches) is ideal voltage source E_{V1}, so $V_{n,V1} = E_V$ and space $\overline{V_{n,F}}$ is of dimension 1 (line). Branch *2* (cotree branch) is admittance branch, so $I_{m,F} = I_{m,Y}$ and space $\overline{I_{m,F}}$ is of dimension 1 [line (a_2)]. It means that the dimension of network space $\overline{K_{I,V,F}}$ is equal $b - f'' = 2$ and is shown as the two-dimensional lattice plane in Fig. 8.2.

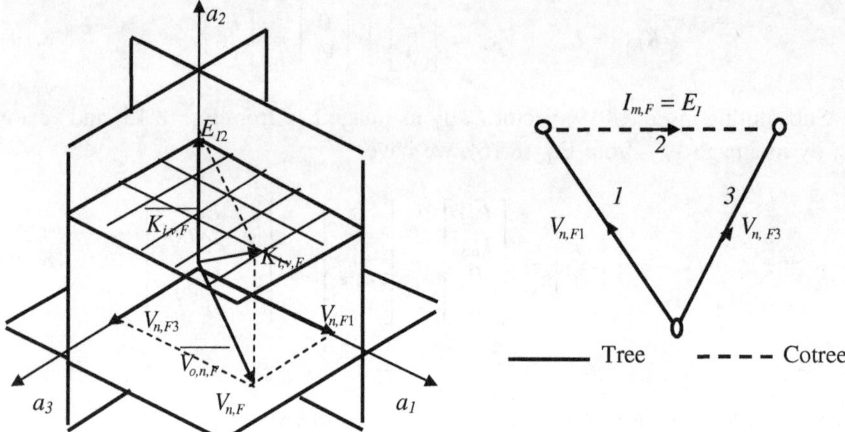

Fig. 8.1 Illustration of space $\overline{K_{I,V,F}}$ in case if branch *2* is ideal current source

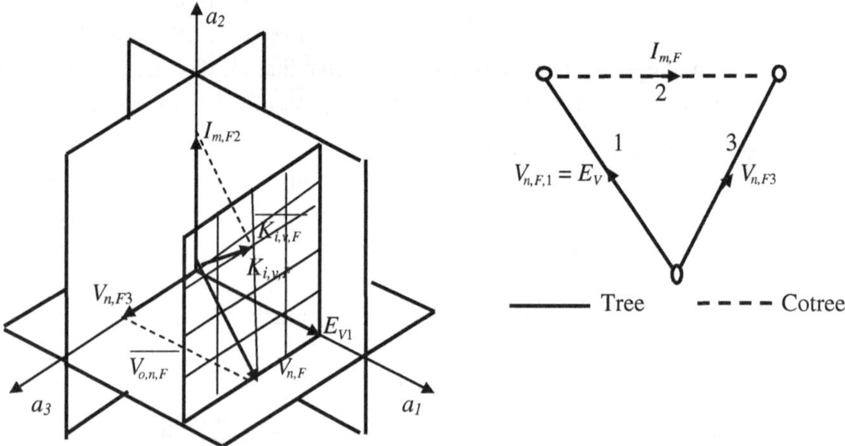

Fig. 8.2 Illustration of space $\overline{K_{I,V,F}}$ in case if branch *1* is ideal voltage source

3. If the current and voltage sources exist, then in Eq. (8.15) matrices $Y_{m,F}$ and $Z_{n,F}$ are restricted. It means that the network-current–voltage state vectors $K_{I,v,F}$ include $f' + f''$ constant elements. The space $\overline{I_{m,F}}$ is of dimension zero and space $\overline{V_{n,F}}$ is of dimension 1 So the space $\overline{K_{I,V,F}}$ is of dimension $b - (f' + f'') = 1$. It can be illustrated if lattice linear spaces in Figs. 8.1 and 8.2 are drawn together in one linear space (not shown), in which intersection of lattice planes forms a line. It means that space $\overline{K_{I,V,F}}$ is one-dimensional.

8.4 Formulation of Ohm's Law Using Matrix T

The network system model developed in previous sections was derived in the form of the current and voltage state vectors $I_{m,F}$ and $V_{n,F}$. In the practical point of view, it is of advantage to derive this model in the form of current and voltage vectors I and V. In this section, such model is derived using topological matrix T.

Consider an electrical, b-branch network system in which current sources are in cotree and voltage sources are in tree and each branch which includes the source is separated into two branches: first, branch which has ideal current or voltage source, and second, branch which has current–voltage function. The Ohm's law can be formulated in matrix form as follows:

$$F_{m,n}\, K_{I,V,F} = K_{m,n} \tag{8.21}$$

where

$F_{m,n}$ The *matrix of branch parameters* called the *network parameters matrix.*

$$F_{m,n} = \begin{bmatrix} 1^{f'} & & & \\ & Z_{m,y} & & \\ & & 1^{f''} & \\ & & & Y_{n,z} \end{bmatrix} \tag{8.22}$$

Assuming that there are no mutual couplings in the network, the matrix $F_{m,n}$ consists of two types of diagonal submatrices: the unit submatrices indicating ideal current and voltage sources E_I and E_V, respectively, and the current–voltage function matrices of impedance and admittance types. Note that, different as in previous sections, the passive branch parameters of cotree branches are impedances and passive branch parameters of tree branches are admittances. Generally, the ordering of submatrices in matrix $F_{m,n}$ is optional. Let us assume that the ordering of submatrices in matrix $F_{m,n}$ is as follows:

- $1^{f'}$ Unit matrix indicating the ideal current sources in cotree, of order $f' \times f$ (f' is a number of current sources).
- $Z_{m,y}$ submatrix of impedance branches in cotree, of order $m - f'$ ($m - f'$ is a number of impedance branches).
- $1^{f''}$ Unit matrix indicating the voltage sources, of order f'' (f'' is a number of voltage sources).
- $Y_{n,z}$ Submatrix of tree admittance branches of order $n - f''$ ($n - f''$ is a number of admittance branches).

Using vector $K_{i,v,F}$ in the form (see Eq. 8.20):

$$K_{i,v,F} = \begin{bmatrix} I_{m,I} \\ I_{m,y} \\ V_{n,V} \\ V_{n,z} \end{bmatrix} \tag{8.23}$$

where

$I_{m,I} = E_I$ vector of ideal current sources in cotree, of order f'.
$I_{m,y}$ vector of branch impedances in cotree, of order $m - f'$.
$V_{n,V} = E_V$ vector of ideal voltage sources in tree, of order f''.
$V_{n,z}$ vector of branch admittances in tree, of order $n - f''$.

Vector $K_{m,n}$ in Eq. (8.21) is a current–voltage vector; arranged according to the ordering of matrix $F_{m,n}$.

$$K_{m,n} = F_{m,n} K_{i,v,F} = \begin{bmatrix} E_I \\ V_{m,y} \\ E_V \\ I_{m,z} \end{bmatrix} \tag{8.24}$$

where

$V_{m,y}$ voltage vector of impedance branches in a cotree, of order m_z.
$I_{n,z}$ current vector of admittance branches in a tree, of order n_y.

Eq. (8.24) is the Ohm's law only. In order to take into account the Kirchhoff's laws, let us use topological transformation matrix T (see Chap. 6; Eq. 6.9):

$$K_{i,v,F} = T^{-1}\, K \tag{8.25}$$

Substituting Eq. (8.25) into (8.26), we have

$$K_{m,n} = F_{m,n} T^{-1} K \tag{8.26}$$

where $K = I + V$

If the topological matrix T is known and if parameters of passive and active branches are nonzero values, then matrix $F_{m,n}$ is non-singular.

Denoting:

$$R = F_{m,n}\, T^{-1} \tag{8.27}$$

The Ohm's law using matrix T is as follows:

$$R\ K = K_{m,n} \tag{8.28}$$

8.5 Matrices of System Admittance and System Impedance

Substituting equation $K = I + V$ into Eq. (8.28), we have

$$R\,I + R\,V = K_{m,n} \tag{8.29}$$

Note that if writing the matrix equation (8.29) in the form of a set of b scalar equations, then on the left- and right-hand sides of equations there are the elements containing the same variables (branch currents and branch voltages). Transferring this variables from the right- to the left-hand side of Eq. (8.29) and adding accordingly, we have

$$R_I\,I + R_V\,V = 0 \tag{8.30}$$

where

R_I matrix R after transferring branch currents.
R_V matrix R after transferring branch voltages.

Equation (8.30) can be interpreted in two ways either the current vector I as a function of voltage vector V or the voltage vector V as a function of current vector I.

$$I = -R_1^{-1} R_2\,V \tag{8.31}$$

$$V = -R_2^{-1} R_1\,I \tag{8.32}$$

Denoting:

$$Y_S = -R_1^{-1} R_2 \tag{8.33}$$

$$Z_S = -R_2^{-1} R_1\,V \tag{8.34}$$

where

Y_S can be called *the matrix of network system admittance.*
Z_S can be called *the matrix of network system impedance.*

Finally Eqs. (8.31 and 8.32) are:

$$I = Y_S\,V \tag{8.35}$$

$$V = Z_S\,I \tag{8.36}$$

The above equations are the *network system equations* relating b-dimensional vectors: current vector I and voltage vector V of a network system. Eqs. (8.35 and 8.36) are a non-conventional formulation of the Ohm's laws of network system, which include the branches which are current and/or voltage sources. Note that each real network, called in this book *the network system*, must include at least one current or voltage source. Without any source, the network is empty ($I = 0$, $V = 0$) and the classical network Ohm's law in such network does not exist. The classical network Ohms law can be used to the parts of network. The Ohms law of network system, derived above, refers to the whole network system which includes at least one source.

Note the following special properties of network system matrices Y_S and Z_S

1. The network system matrices Y_S, Z_S and equations $I = Y_S V, V = Z_S I$ fulfill the fundamental equation:

$$Y_S\, Z_S = \; R_I^{-1}\, R_V\, R_V^{-1}\, R_I = 1 \tag{8.37}$$

2. The matrices Y_S, Z_S, and equations $I = \; Y_S\, V,\; V = \; Z_S\, I$ are the analogy to the classical admittance and impedance matrices and classical Ohm's equations; however, they cannot be identified with them. Note the important difference between the classical matrices Y, Z, and the system matrices Y_S and Z_S. The elements of classical matrices are pure branch parameters only, independent on network as a whole; the elements of system matrices are network system parameters, dependent on network as a whole.
3. Matrices Y_S and Z_S are not diagonal (even if matrix $F_{m,n}$ is diagonal); so they are analogy to the self- and mutual branch couplings of a network, but they cannot be identified with them. Elements of matrices Y_S, Z_S are system self- and system mutual admittances and impedances, which numerically differ from classical branch admittances/impedances. They are the whole network system quantities.
4. Elements of matrices Y_S and Z_S are network parameters of branches which not only takes into account branch admittances/impedances but also takes into account ideal current and voltage sources, which are not admittances/ impedances. They are *system* parameters characterizing actual state of network system.
5. As it comes from Eqs. (8.33 and 8.34), the existence of system admittance Y_S and impedance Z_S depend on the non-singularity of matrices R_I and R_V. Analysis of a simple network example, which are given in Chap. 14 shows that these matrices may be non-singular or singular depending on the number and configuration of ideal current and voltage sources in a network (see Chap. 14). However, it is difficult to formulate the mathematical conditions of existence Y_S and Z_S because there may be some other reasons of singularity of matrices R_I and R_V.

Chapter 9
General Comments to Part I

The book presents a modern and non-conventional approach to the theoretical analysis and solution of electrical networks systems. The generally unknown interrelations of network quantities and the formulations of the fundamental network laws are derived. Some possibilities of its application in network analysis and solution are given. Formulations are based on the original mathematical (linear algebra) model of a network. It was shown that modelling network graph in the form of a topological incidence matrix T makes possible derivation of various formulations of network laws and is widening the network analysis. Various matrix equations of voltage and current relations and general network solution equations are derived. The algebraic model of a network topology leads to some unexpected results. It was proved that summation of current and voltage values is mathematically sensible. The Ohm's law of network system, which takes into account ideal current and voltage sources are derived and illustrated.

Commenting generally, it seems to me that what was presented in this part of book may lead to the following general remarks (conclusions), namely:

- First: that the network system is "the totality which is not a sum of its parts." The network system has its own character and features; has its own mathematical quantities: vectors, matrices, linear spaces; and has, e.g., its own Ohm's law, etc.
- Second: It seems to us that the notions of electrical current and electrical voltage are not entirely different physical quantities. Branch current and branch voltage are of course the different physical quantities, but the network current state and the network voltage state may have more in common. The facts that the current state and voltage state vectors are orthogonal; that the notion of common, current–voltage vector of network has physical meaning; that there exists the topological interdependency between currents and voltages; that the vector of sums of a branch current and a branch voltage have the mathematical meaning;

A. Kłos, *Mathematical Models of Electrical Network Systems*,
Lecture Notes in Electrical Engineering 412,
DOI 10.1007/978-3-319-52178-7_9

and that the Kirchhoff's and Ohm's laws can be united in a one network law all these are interesting, worth of notice and need the further theoretical investigations, which may lead to the development of various practical applications. Generally, the theory of network systems needs further investigations, taking into account nonlinear functions and dynamic states of networks.

An overall aim of the work described in the first part of book was to develop a mathematical model of network topology and to derive the topological transformation matrix T which appears to be a useful tool of electrical network analysis and solution.

The algebraic model of electrical network presented in the first part, independent of any practical applications, is widening the theoretical background and may serve to the further development of network knowledge.

Note that in the above presented algebraic network, the primary notion (not to be mathematically defined) is the network branch only. Different than in graph theory, the node is defined as a set of branches. Practical application of the above formulas does not need introducing the notion of slack node.

9.1 Part II Application Examples

In the second part of the book, some applications of algebraic model of network system are presented. Taking into account that the knowledge of electrical networks is used in many fields of science and technology, the algebraic model may have various applications. It is, of course, not possible even mention all of them in this part of book. However, some applications concerning power system networks can be presented in what follows. The recently used methodology of the electrical network analysis and solution is based on rather simple mathematical model and was developed mainly for needs in the field of power system networks. The aim of this part of the book was widening the practical analysis and solution of such networks, by using the algebraic model presented in the first part.

So far it was not necessary to distinguish the quantities used in model, into known input data and the unknown to be found quantities. The numerical state of variables is irrelevant, since the object of theory is only to find the mathematical model, which reflects all the possible relationships between the quantities. In this part, the subdivision into input data and unknown variables is necessary. The applications are divided into the following:

- Applications of the non-conventional formulations of Kirchhoff's laws given in Chap. 7 to the practical network system analysis.
- Applications of the algebraic model of network system, derived in the first part of book, to the widening of methodology of network solution methods.

Part II
Application Examples

In the second part of book, some applications of algebraic model of network system are presented. Taking into account that the knowledge of electrical networks is used in many fields of science and technology, the algebraic model may have various applications. It is, of course, not possible even mentioning all of them in this part of book. However, some applications concerning power system networks can be presented in what follows. The recently used methodology of the electrical network analysis and solution is based on rather simple mathematical model and was developed mainly for needs in the field of power system networks. An aim of this part of book is widening the practical analysis and solution of such networks, by using the algebraic model presented in first part.

So far, it was not necessary to distinguish the quantities used in model, into known input data and the unknown to be found quantities. The numerical state of variables is irrelevant, since the object of theory is only to find the mathematical model, which reflects all the possible relationships between the quantities. In this part, the subdivision into input data and unknown variables is necessary. The applications are divided into:

- Applications of the non-conventional formulations of Kirchhoff's laws given in Chap. 7 to the practical network system analysis and
- Applications of the algebraic model of network system, derived in the first part of book, to the widening of methodology of network solution methods.

Chapter 10
Applications of *b*-Dimensional Formulation of Kirchhoff's Laws

10.1 Basic *b*-Dimensional Formulation

The non-conventional formulations of Kirchhoff's laws, given in the first part of book (see Chap. 7), may have various applications; some of them are given in this chapter.

In this section, the application of the basic b-dimensional formulation of Kirchhoff's law is given [see Eqs. (7.7) and (7.8)]. In Chaps. 4 and 5, various current relations and voltage relations are derived in the form of m- and n-order equations. The question arises what these equations have to do with b-dimensional current and voltage equations.

Consider the current equation. It is well known that the tree currents can be found from known cotree currents using Eq. (4.6) ($I_n = -C_m I_m$). In the classical network analysis, the inverse relations of Eq. (4.6) do not exist because matrix C_m is singular. However, using algebraic model of network, for some kinds of networks, such relations do exist and can be derived in the matrix form using the b-dimensional Kirchhoff's current law in the form of Eq. (7.7) ($A_I I = 0$). In Eq. (7.7), matrix A_I can be subdivided into four submatrices:

$$\begin{bmatrix} A_{I,1} & A_{I,2} \\ A_{I,3} & A_{I,4} \end{bmatrix} \begin{bmatrix} I_m \\ I_n \end{bmatrix} = 0 \tag{10.1}$$

where

$A_{I,1}$ is a non-singular matrix of order $m \times m$,
$A_{I,2}$ matrix of order $m \times n$,
$A_{I,3}$ matrix of order $n \times m$,
$A_{I,4}$ a non-singular matrix of order $n \times n$.

A. Kłos, *Mathematical Models of Electrical Network Systems*,
Lecture Notes in Electrical Engineering 412,
DOI 10.1007/978-3-319-52178-7_10

Selecting the first row of Eq. (10.1), we have

$$A_{I,1}I_m + A_{I,2}I_n = 0 \tag{10.2}$$

For any structure of network topology, if the number of cotree branches is less than a number of tree branches, $m < n$, then the matrix $A_{I,1}$ is non-singular and of order m, and cotree current vector can be found as a function of tree current vector from Eq. (10.2).

$$I_m = -A_{I,1}^{-1}A_{I,2}I_n \tag{10.3}$$

where matrix $A_{I,1}^{-1}A_{I,2}$ is not incidence (1, 0, −1) matrix.

It means that the relation I_m as a function of I_n does exist, but for special kind of networks only. Note that the second row of Eq. (10.3) is the classical relation $I_n = -C_m I_m$ and matrix $A_{I,4}^{-1}\,A_{I,3}$ is equal to the incidence matrix C_m.

Consider the voltage equation. It is well known that the cotree voltages can be found from known tree voltages using Eq. (5.6) ($V_m = -L_n V_n$). In the classical network analysis, the inverse relations of Eq. (5.6) do not exist because matrix L_n is singular. However, using algebraic model of network, for some kinds of networks, such relations do exist and can be derived in the matrix form using the b-dimensional Kirchhoff's voltage law in the form of Eq. (7.8) ($A_V V = 0$). In Eq. (7.9), matrix A_V can be subdivided into four submatrices:

$$\begin{bmatrix} A_{V,1} & A_{V,2} \\ A_{V,3} & A_{V,4} \end{bmatrix} \begin{bmatrix} V_m \\ V_n \end{bmatrix} \tag{10.4}$$

where

- $A_{V,1}$ is a non-singular matrix of order $m \times m$,
- $A_{V,2}$ matrix of order $m \times n$,
- $A_{V,3}$ matrix of order $n \times m$, and
- $A_{V,4}$ a non-singular matrix of order $n \times n$.

Selecting the second row of Eq. (10.4), we have

$$A_{V,3}V_m + A_{V,4}V_n = 0 \tag{10.5}$$

For any structure of network topology, if the number of cotree branches is greater than a number of tree branches, $m > n$, then the matrix $A_{V,4}$ is non-singular and of order n, and the tree voltage vector can be found as a function of cotree voltage vector from Eq. (10.5).

$$V_n = -A_{V,4}^{-1}A_{V,3}V_m \tag{10.6}$$

It means that the relation V_n as a function of V_m does exist, but for special kind of networks only. Note that the first row of Eq. (10.4) is the classical relation: $V_m = -L_n V_n$ and the matrix $A_{V,4}^{-1} A_{V,3}$ is equal to the incidence matrix $-L_m$.

The equations derived above, and some others, non-common not known relations, which can be derived by subdividing the b-dimensional formulation of Kirchhoff's law, may be useful in the network analysis.

10.2 Pseudo-Unit Formulation

Consider the pseudo-unit formulation of Kirchhoff's laws, derived in Chap. 7 [see Eqs. (7.13) and (7.14)]. Starting from these equations, the pseudo-unit matrices B_I and B_V are derived which relate every branch current with the remaining branch currents and every branch voltage with the remaining branch voltages.

$$I = B_I I \tag{10.7}$$

$$V = B_V V \tag{10.8}$$

Note that the matrices B_I and B_V can be found knowing topological (tree/cotree) structure of network only. The matrices B_I and B_V are non-singular and have interesting properties, which may be useful in network analysis. In this section, an example of practical application of the pseudo-unit current equation $I = B_I I$ is presented.

It is intuitively known that any changing of a current or a voltage in any branch of network system will have certain influence on the currents and voltages of the other network branches. The Eqs. (10.7) and (10.8) enable the quantitative and qualitative analyses of this influence. Note that changing a steady state of a real network, namely changing the branch current i_j by constant value Δi_j, generate the changing of all branch currents, among them changing current of branch i_j. In order to keep the value Δi_j constant, the branch j must be the regulated current source. The problem of finding the numerical values of branch currents after any change in one branch can be solved using the following iteration procedure:

Let us write the pseudo-unit current equation $I = B_I I$ as follows:

$$I_a = B_I I_b \tag{10.9}$$

For any network current vector $I = I_a = I_b$, Eq. (10.8) must be fulfilled. If in the current vector I_b the element $i_{b,j}$ (current of branch j) is changed by adding arbitrary value $\Delta i_{b,j}$, then Eq. (10.8) is not fulfilled. However, using the simple iteration process, one can find a new, unique vector $I'_a = I'_b = I'$ which fulfills the Eq. (10.8). The iteration process is as follows:

1. Choose any starting point current vectors $I = I_a = I_b$; chose the current of branch j to be changed $i_{j,a} = i_{j,b} = i_j$.
2. Choose the change $\Delta i_{j,b}$ of branch current $i_{j,b}$ and find the new, constant current of branch j $i^k_{j,b} = i_{j,b} + \Delta i_{j,b}$.
3. Change the current vector I_b by substituting $i_{j,b} = i^k_{j,b}$.
4. Find new vectors $I'_b = I'_a$ using Eq. (10.8).
5. If $i'_{j,b} = i^k_{j,b}$, then go to 7.
6. If $i'_{j,b}$ is not equal to $i^k_{j,b}$, then $I_b = I'_b$ and go to 3.
7. End of iteration process. It means that the pseudo-unit current Eq. (10.8) is fulfilled and the starting point vectors $I = I_a = I_b$ are the new current vectors $I'_a = I'_b = I'$.

Note that the reverse iteration process leads to the starting point current vector $I = I_a = I_b$.

Comparing the new branch currents with the starting point branch currents makes possible finding of how changing of branch current $i_{b,j}$ influences on the other branch currents in network. It is possible to do the accurate analysis of branch currents interrelations.

In what follows such analysis is illustrated for a simple example of network in Fig. 10.1.

Consider a network in which the number of cotree branches is greater than the number of tree branches. An example of the network graph, the topology of

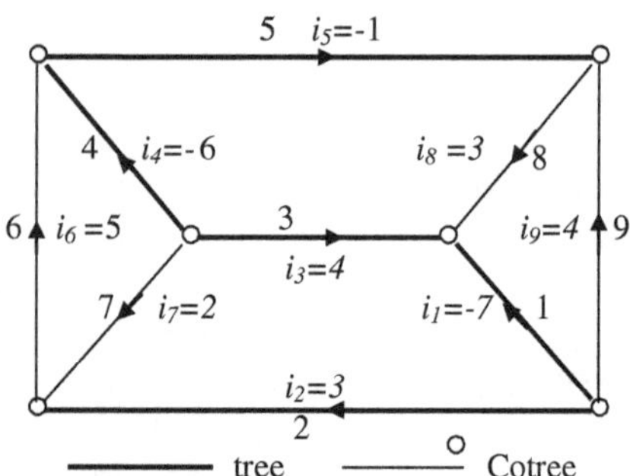

B_l =

	6	7	8	9	1	2	3	4	5
6	0.000	0.500	0.125	-0.250	-0.125	0375	0.000	-0.500	0.375
7	0.500	0.000	0.125	0.125	0.250	-0.375	-0.375	-0.500	0.000
8	0.125	0.125	0.000	0.500	-0.500	0.000	-0.375	0.250	0.375
9	-0.250	0.125.	0.500	0.000	-0.500	-0.375	0.000	-0.125	-0.375
1	-0.125	0.250	-0.500	-0.500	0.000	-0.375	-0.375	0.125	0.000
2	0.333	-0.333	0.000	-0.333	-0.333	0.000	0.333	0.000	0.333
3	0.000	-0.333	-0.333	0.000	-0.333	0.333	0.000	-0.333	-0.333
4	-0.500	-0.500	0.250	-0.125	0.125	0.000	-0.375	0.000	0.375
5	0.333	0.000	0.333	-0.333	0.000	0.333	-0.333	0.333	0.000

Fig. 10.1 Example of the pseudo-unit matrix B_l of a simple network

network, the starting point branch currents and matrix B_I is shown in Fig. 10.1. Suppose that branch currents can be regulated.

The analysis of elements of matrix leads to the following conclusions:

- If the non-diagonal elements $b_{j,k}$ are equal to zero (e.g., element $b_{j,k} = b_{j,5}$), then it means that current of branch j does not depend on the current changes of branch k and opposite.
- Greater/lower values of non-diagonal elements $b_{j,k}$ mean greater/lower changing of current of branch j on the current of branch k and opposite.

Using the above matrix B_I, the two cases of numerical calculations (using simple computer program) can be solved:

In the first case, the current of branch 1 is changed by adding number 1, $\Delta i_{b,1} = 1$. The new, constant current of branch 1, $\Delta i^k_{b,1} = -6.000$.

In the second case, the current of branch 9 is changed by adding number −2, $i_{b,9} = -2$. The new current of branch 9 $i^k_{b,9} = 2.000$.

The results of case 1 are as follows:

Branch currents	i_6	i_7	i_8	i_9	i_1	i_2	i_3	i_4	i_5
Starting vector $I = I_a = I_b$	5.000	2.000	3.000	4.000	−7.000	3.000	4.000	−6.000	−1.000
New vector $I' = I'_a = I'_b$	4.865	2.270	2.407	3.407	−6.000	2.593	3.593	−5.865	−1.000
Difference $I' - I$	−0.135	0.270	−0.593	−0.593	1.000	−0.407	−0.407	0.135	0.000
Difference %	−2.7	11	−19.7	−14.8	14.3	−13.5	−10.2	2.0	0

The results of case 2 are as follows:

Branch currents	i_6	i_7	i_8	i_9	i_1	i_2	i_3	i_4	i_5
Starting vector $I = I_a = I_b$	5.000	2.000	3.000	4.000	−7.000	3.000	4.000	−6.000	−1.000
New vector $I' = I'_a = I'_b$	5.544	1.728	1.812	2.000	−5.819	3.813	4.000	−5.728	−0.187
Difference	0.544	−0.272	1.188	−2.000	1.181	0.813	4.000	0.272	0.813
Difference %	10.9	13.6	39.6	50.0	16.8	27.0	0	4.5	18.7

In the first case, 14.3% change of current in branch 1 leads to the highest, that is, equals to 19.7%, reduction of current in branch 8; the reductions of current from 2 to 14.8% in other branches, and not any change of current in branch 5.

In the second case, 50% change of current in branch 9 leads to the highest, that is, equals 39.6%, reduction of current in branch 8; reductions from 4.5 to 27% in other branches, and not any change of current in branch 3.

Generally, using the pseudo-unit formulation of Kirchhoff's laws, the qualitative analysis and control of interdependences of branch current flow and branch voltage values in electrical network branches can be conducted. Additionally, if the branch voltages and branch impedances in the starting point are known, then one can analyze the influence of any changes of branch currents on the branch voltages (and powers) can be done.

Chapter 11
Network Solution Method Using Algebraic Network Model

11.1 Derivation of Method Using Matrix *T*

Using the algebraic model of network derived in Part I, the general method of linear network solution is derived in this section. Suppose that the graph of b-branch network is known, and the topology is unrestricted, except that a tree includes voltage sources and a cotree includes current sources. The network branches are numbered from 1 to b (the first m are cotree branches, and the last n are tree branches). Generally, the ordering of branches in the cotree and tree is optional. Not loosing generality, the ordering in vectors and matrices of order $b = m + n$ is assumed as follows:

- First m_i branches are cotree current sources, $m_i \leq m$.
- Next $m - m_i$ branches are cotree admittances.
- Next n_v branches are tree voltage sources, $n_v \leq n$.
- Last $n - n_v$ branches are tree impedances.

In what follows, all vectors and matrices are of order b.

Starting from the b-dimensional formulation of the Kirchhoff's laws [see Chap. 4, Eq. (4.11) and Chap. 5, Eq. (5.10)].

$$I = TI_{m,o} \tag{11.1}$$

$$V = TV_{o,n} \tag{11.2}$$

where

I and V	current and voltage vectors.
$I_{m,o}$ and $V_{o,n}$	current state vector and voltage state vector [see (4.12), (5.11)],
T	topological matrix.

The vectors $I_{m,o}$ and $V_{o,n}$ include the current and voltage sources.

A. Kłos, *Mathematical Models of Electrical Network Systems*,
Lecture Notes in Electrical Engineering 412,
DOI 10.1007/978-3-319-52178-7_11

$$I_{m,o} = E_I + I_Y \tag{11.3}$$

$$V_{o,n} = E_V + V_Z \tag{11.4}$$

where

E_I current vector of cotree current sources
I_Y current vector of cotree admittance branches
E_V voltage vector of tree voltage sources
V_z voltage vector of tree impedance branches.

Introducing the elementary Ohm's law:

$$I_Y = Y_m V \tag{11.5}$$

$$V_Z = Z_n I \tag{11.6}$$

where

Y_m matrix of cotree branch admittances.
Z_n matrix of tree branch impedances.

Substituting Eq. (11.1) into (11.6) and Eq. (11.2) into (11.5), we have:

$$I_Y = Y_m T V_{o,n} \tag{11.7}$$

$$V_Z = Z_n T I_{m,o} \tag{11.8}$$

Substituting Eq. (11.9) into (11.8) and Eq. (11.10) to (11.7), we have:

$$I_{m,o} = E_I + Y_m T V_{o,n} \tag{11.9}$$

$$V_{o,n} = E_V + Z_n T I_{m,o} \tag{11.10}$$

Substituting Eq. (11.9) to (11.8) and Eq. (11.10) to (11.7), we have:

$$I_{m,o} = E_I + Y_m T (E_V + Z_n T I_{m,o}) \tag{11.11}$$

$$V_{o,n} = E_V + Z_n T (E_I + Y_m T V_{o,n}) \tag{11.12}$$

After doing simple algebraic operations, the current and voltage state vectors $I_{m,o}$ and $V_{o,n}$ are as follows:

$$I_{m,o} = (1 - Y_m T Z_n T)^{-1} (E_I + Y_m T E_V) \tag{11.13}$$

$$V_{o,n} = (1 - Z_n T Y_m T)^{-1} (E_V + Z_n T E_I) \tag{11.14}$$

Substituting Eq. (11.13) to Eq. (4.12) and Eq. (11.14) to Eq. (5.12) we have the final equations of linear network solution using matrix T.

$$I = T(1 - Y_m T Z_n T)^{-1}(E_I + Y_m T E_V) \tag{11.15}$$

$$V = T(1 - Z_n T Y_m T)^{-1}(E_V + Z_n T E_I) \tag{11.16}$$

Equations (11.13)–(11.16) relate all network quantities, and they can be used to the solution of these quantities and to the solvability analysis depending on the given data.

11.2 Classical Network Solution Methods

Equations (11.15) and (11.16) are generalization of classical nodal and mesh methods of power system network solution. After assuming some simplifications (done in classical analysis), these equations can be reduced to the well-known node method and mesh method equations, as follows.

The node method is based on the following assumptions. The topology in classical network analysis is usually assumed as follows: All node-to-earth branches are tree branches, and all node-to-node branches are cotree branches. The voltage vector of tree branches V_n and the admittance vector Y_m of cotree branches are known constant values. The unknown and to be found are the vector of tree currents I_n. It means that in Eq. (11.15), V_n is known voltage source vector $E_V = V_n$ and the vector of current sources $E_I = 0$. The vector of tree impedances $Z_n = 0$ and all cotree branches are known admittances Y_m. Substituting these data to Eq. (11.15) makes this equation much simpler. If $E_I = 0$ and $Z_n = 0$, then $(1 - Y_m T Z_n T)^{-1}$ is equal to unit matrix. So the Eq. (11.15) reduces to the following equation:

$$\begin{bmatrix} I_m \\ I_n \end{bmatrix} = \begin{bmatrix} 1 & C^T \\ -C & 1 \end{bmatrix} \left(\begin{bmatrix} Y_m & 0 \\ 0 & Z_n \end{bmatrix} \begin{bmatrix} 1 & C^T \\ -C & 1 \end{bmatrix} \begin{bmatrix} 0 & 0 \\ 0 & E_V \end{bmatrix} \right)$$

Selecting the equation of order n, we have:

$$I_n = -C Y_m C^T V_n \tag{11.17}$$

which is classical node method of network solution.

Analogically, the classical mesh method is based on the following assumptions. The current vector of cotree branches I_m and the vector of tree branch impedances Z_n are known constant values. The unknown and to be found are the vector of cotree voltages V_m. It means that in Eq. (11.16), the current vector of cotree branches I_m is equal to E_I and the voltage source vector $E_V = 0$. The cotree admittances $Y_m = 0$. Substituting these data to Eq. (11.16) makes this equation much simpler. If $E_V = 0$ and $Y_m = 0$, then $(1 - Z_n\, T\, Y_m\, T)^{-1}$ is equal to unit matrix. Equation (11.16) reduces to the following equation:

$$\begin{bmatrix} V_m \\ V_n \end{bmatrix} = \begin{bmatrix} 1 & C^T \\ -C & 1 \end{bmatrix} \left(\begin{bmatrix} 0 & 0 \\ 0 & Z_n \end{bmatrix} \begin{bmatrix} 1 & C^T \\ -C & 1 \end{bmatrix} \begin{bmatrix} E_I & 0 \\ 0 & 0 \end{bmatrix} \right)$$

Selecting the equation of order m, we have:

$$V_m = -C^T Y_m C^T I_m \tag{11.18}$$

which is the classical mesh method of network solution.

The general method of network solution and using various kinds of network topology make possible the derivation of methods and solution problems in cases of various data given. The examples of such methods are given in the next sections.

Chapter 12
Method of Current and Voltage Sensitivity Analysis

12.1 Derivation of Method

It is intuitively known that the current or voltage values of a network branch depend on the current or voltage values of all other branches in the network (practically may depend on certain part of network only). In particular applications, it depends on some control variables which can be identified with current and voltage sources. Generally, any branch current or branch voltage, if it is controllable, can be conventionally treated as a current or voltage source. The method derived in this chapter provided the information on how the current or voltage state of network depends on the current and voltage sources.

Let us start from generalized network solution Eqs. (11.15) and (11.16).

$$I = T(1 - Y_m T Z_n T)^{-1}(E_I + Y_m T E_V)$$

$$V = T(1 - Z_n T Y_m T)^{-1}(E_V + Z_n T E_I)$$

Denoting:

$$F_m = T(1 - Y_m T Z_n T)^{-1} \quad (12.1)$$

$$F_n = T(1 - Z_n T Y_m T)^{-1} \quad (12.2)$$

$$F_{m,y} = T(1 - Y_m T Z_n T)^{-1} Y_m T \quad (12.3)$$

$$F_{n,z} = T(1 - Z_n T Y_m T)^{-1} Z_n T \quad (12.4)$$

Using the above notation, the relation between (11.15) and (11.16) can be written as a function of current and voltage sources.

A. Kłos, *Mathematical Models of Electrical Network Systems*,
Lecture Notes in Electrical Engineering 412,
DOI 10.1007/978-3-319-52178-7_12

$$I = F_m E_I + F_{m,y} E_V \tag{12.5}$$

$$V = F_n E_V + F_{n,z} E_I \tag{12.6}$$

where

E_I vector of ideal current sources, of order b, which has m_i nonzero elements, which are ideal current sources and

E_V vector of ideal voltage sources, of order b, which has m_v nonzero elements, which are ideal voltage sources.

After selecting in Eq. (12.5) the m_i columns of matrix F_m, which relate to the current source vector E_I, and selecting in Eq. (12.6) the m_v columns of matrix F_m, which relate to the voltage source vector E_I, the Eqs. (12.5) and (12.6) can be written as follows:

$$\begin{bmatrix} I_m \\ I_n \end{bmatrix} = \begin{bmatrix} F_m^1 & F_m^2 \\ F_m^3 & F_m^4 \end{bmatrix} \begin{bmatrix} E_I \\ 0 \end{bmatrix} + \begin{bmatrix} F_{m,y}^1 & F_{m,y}^2 \\ F_{m,y}^3 & F_{m,y}^4 \end{bmatrix} \begin{bmatrix} 0 \\ E_V \end{bmatrix} = \begin{bmatrix} F_m^1 & E_I \\ F_m^3 & E_I \end{bmatrix} + \begin{bmatrix} F_{m,y}^2 & E_V \\ F_{m,y}^4 & E_V \end{bmatrix} \tag{12.7}$$

$$\begin{bmatrix} V_m \\ V_n \end{bmatrix} = \begin{bmatrix} F_n^1 & F_n^2 \\ F_n^3 & F_n^4 \end{bmatrix} \begin{bmatrix} 0 \\ E_V \end{bmatrix} + \begin{bmatrix} F_{n,z}^1 & F_{n,z}^2 \\ F_{n,z}^3 & F_{n,z}^4 \end{bmatrix} \begin{bmatrix} E_I \\ 0 \end{bmatrix} = \begin{bmatrix} F_n^2 & E_V \\ F_n^4 & E_V \end{bmatrix} + \begin{bmatrix} F_{n.z}^1 & E_I \\ F_{n,z}^3 & E_I \end{bmatrix} \tag{12.8}$$

where matrices with upper index 1, 2, 3, and 4 are submatrices selected from matrices $F_m, F_{m,y}$ and from matrices $F_n, F_{n,z}$. These equations are of order b. Reducing the order of Eqs. (12.7) and (12.8) and equations E_I and E_V, the solution equations are as follows:

$$I = \begin{bmatrix} F_m^1 & F_{m,y}^2 \\ F_m^3 & F_{m,y}^4 \end{bmatrix} \begin{bmatrix} E_I \\ E_V \end{bmatrix} \tag{12.9}$$

$$V = \begin{bmatrix} F_n^2 & F_{n,z}^1 \\ F_n^4 & F_{n,z}^3 \end{bmatrix} \begin{bmatrix} E_I \\ E_v \end{bmatrix} \tag{12.10}$$

Finally, the Eqs. (12.9) and (12.10), written in the simplest form, are as follows:

$$I = H_I E \tag{12.11}$$

$$V = H_V E \tag{12.12}$$

where

E vector of current and voltage sources of order $m_i + n_{v,}$

H_I matrix of order $b \times (m_i + n_v)$ constructed from columns of matrices $F_m, F_{m,y}$, which belong to the vector E, and

H_V matrix of order $b \times (m_i + n_v)$ constructed from columns of matrices $F_{m,z}, F_n$, which belong to the vector E.

Each element j, k of matrix H_I is the sensitivity coefficient of current of branch j versus one of the sources (voltage source $E_{V,k}$ or current source $E_{I,k}$) of branch k. It describes how much branch voltage i_j depends on the value of current or voltage source. In other words, the multiplication of sensitivity coefficient j, k of matrix H_I by element k of matrix E ($E_{I,k}$ or $E_{V,k}$) is a part of branch current i_j. The numerical values of sensitivity coefficients in jth row of matrix H_I designate the numerical influence of current and voltage sources on the value of branch voltage i_j.

Analogically, each element j, k of matrix H_V is the sensitivity coefficient of voltage of branch j versus one of the sources (voltage source $E_{V,k}$ or current source $E_{I,k}$) of branch k. It describes how much branch voltage j depends on the value of current or voltage source. In other words, the multiplication of sensitivity coefficient j, k of matrix H_V by element k of matrix E ($E_{I,k}$ or $E_{V,k}$) is a part of branch voltage v_j. The numerical values of sensitivity coefficients in jth row of matrix H_V designate the numerical influence of current and voltage sources on the value of voltage V_j.

Note that the elements of matrices H_I and H_V depend on the passive branch parameters and that the sources are controlled ideal current and ideal voltage sources.

12.2 Numerical Example

Equations (12.11) and (12.12) are illustrated on a simple example of power system network in Fig. 12.1. Branch 4 is the current (power) input (power station); it is also a slack-point branch and voltage source. Branches 1, 2, and 3 are current outputs (loads) and also current sources. Branches 5, 6, 7, 8, and 9 are load transmission lines.

Using passive branch parameters, the numerical results are given in Fig. 12.1 in the form of equations $I = H_I E$ and $V = H_V E$. Elements of matrices H_I and H_V are the sensitivity coefficients. Analysis shows the well-known and characteristic features of power system networks; e.g., that current flow does not depend on voltage level determined by slack node voltage (branch 4). Moreover, analysis leads to some practically useful conclusions.

- Current source $E_{1,2}$ has the greatest influence on transmission current of line 5, has lowest influence on current of transmission line 7, and has, of course, no any influence on current sources $E_{I,1}$ 1 and $E_{I,3}$.
- Current sources have influence on all branch voltages, except on slack-point branch voltage. Current source $E_{1,2}$ has greatest influence on voltage of line 2 and lowest influence on voltage of branch 7.

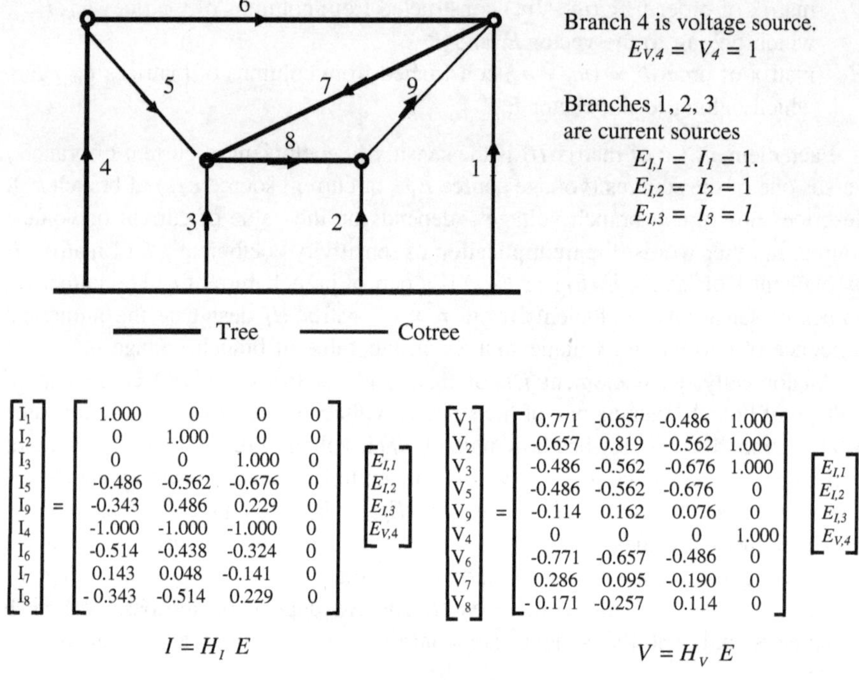

Fig. 12.1 Illustration of sensitivity coefficients of a simple power system network

- Voltage source $E_{V,4}$ has no influence on branch currents but has greatest (nominal) influence on voltages of current source branches 1, 2, and 3.

Generally, the above given method can be used if the value and location of current and voltage sources are to be found, or in problems with needed current–voltage interdependences.

In the power system operation, the method of sensitivity coefficients can be used as a tool of power system security analysis and control. In order to avoid serious disturbances of a whole or part of a power system, the possible full information is needed concerning location and danger of disturbance. The method may serve to provide the quick information of how the branch currents depend on the node currents or transmission line currents (or as a rough estimate of how branch powers depend on the chosen node or transmission line powers) and serve to prevent the emergency situations (e.g., blackouts) in cases of overloads and under- or over-voltages. The sensitivity method may be used in the power system planning problems.

Chapter 13
Method of Arbitrary Input Data

13.1 Description of Arbitrary Input Data

The existing solution methods of linear networks, which have great number of branches, are based on certain assumptions concerning the kind, number, and distribution of given input data, which are strictly defined. They cover majority of the practical network solution problems. However, there are problems in which the form of input data can not be strictly defined. The method of arbitrary input data, presented in this chapter, deals with problems in which the input data are arbitrary, randomly distributed in the network. The method is based on the algebraic model of network derived in the first part of book.

Consider the problem in which the graph of network is known and the input data (values of branch currents, branch voltages, and branch parameters) are arbitrarily distributed on the network branches. It is assumed that for a branch the two or one or no any data may be given. Branches in series are treated as one branch. It is assumed that there are branches with following kinds of data given:

- branch current only—(*i*-branch),
- branch voltage only—(*v*-branch),
- branch parameter (impedance/admittance) only—(*f*-branch),
- branch current and voltage—(*i, v*-branch),
- branch current and branch parameter (impedance/admittance)—(*i, f*-branch),
- branch voltage and branch parameter (impedance/admittance)—(*v, f*-branch),
- branch with not any data given (*o*-branch), and
- It is assumed that the input data can not be changed (e.g., branch with given current and impedance (*i, f*-branch) can not be replaced by a *v*-branch). The input data may be randomly distributed in the network.

It is obvious that the unique solution of such problem may not exist (problem may be inconsistent), so the method of solution must include the solvability analysis. Preliminary analysis of input data, done from the point of view of

A. Kłos, *Mathematical Models of Electrical Network Systems*,
Lecture Notes in Electrical Engineering 412,
DOI 10.1007/978-3-319-52178-7_13

solvability, leads to the condition concerning the distribution of certain kind of data. Problem can be solved if it is possible forming the cotree from some kinds of branch data and forming the tree from some kinds of branch data. In order to use as much as possible input data, it is necessary that the branches with given current must belong to any cotree and the branches with given voltage must belong to any tree (see Chaps. 4 and 5). It may be not possible if tree and cotree are complementary topological structure. Note that if there are branches with given current and voltage (*i, v*-branches), then in case of complementary tree and cotree, they can not be effectively used (current i or voltage v can not be used). Moreover, if there are branches with no data given, then they are useless because they can not belong to a tree and to a cotree. It means that in order to formulate the topology of network, it is necessary to use the two different topologies: one for currents, called *current topology*, represented by *current topological matrix* T_I, and second one for voltages, called *voltage topology*, represented by *voltage topological matrix* T_V. It may be necessary solvability condition. The cotree of current topology denoted by *i-cotree* must include branches with given current input values (*i*-branches, *i, v*-branches, and *i, f*-branches). The tree of voltage topology, denoted by *v-tree*, must include branches with given voltage input values (*v*-branches, *i, v*-branches, and *v, f*-branches). If there are more then m branches with given current or more then n branches with given voltage, the input data are not consistent (may be omitted).

13.2 Derivation of Solution Equations

The solution method of the above-formulated problem can be derived using the algebraic model of electrical network described in Part I. Consider the connected and closed network, with branches numbered from 1 to b. Assuming the two different topologies, *i*-cotree and v-tree should be chosen, the Kirchhoff's laws can be written as follows [see Chaps. 4 and 5, Eqs. (4.12) and (5.12)]:

$$I = T_I I_{m,o} \tag{13.1}$$

$$V = T_V V_{o,n} \tag{13.2}$$

where

- I and V the current and voltage vectors of a network,
- T_i Current topological matrix (*i*-cotree and complementary tree),
- T_V Voltage topological matrix (*v*-tree and complementary cotree),
- $I_{m,o}$ The current state vector of order b; the first m elements are cotree currents, and last n elements are zeros, and
- $V_{o,n}$ The voltage state vector of order b; the first m elements are zeros, and last n elements are tree voltages.

Using the input data, the current cotree vector $I_{m,o}$ can be formulated as the sum of vectors:

$$I_{m,o} = I_i + I_f \tag{13.3}$$

where

- I_i current cotree vector (of order b), which elements include the known branch currents (input data: i-branches, i, v-branches, and i, f-branches).
- I_f current cotree vector which elements can be found using known i-cotree admittances taken from f-branches and f, v-branches and using the unknown voltage state vector $V_{o,n.}$

$$I_f = Y_m T_V V_{o,n} \tag{13.4}$$

where Y_m is the admittance matrix of i-cotree branches.

Analogically, the voltage state vector $I_{m,o}$ can be formulated as the sum of two vectors:

$$V_{o,n} = V_V + V_f \tag{13.5}$$

where

- V_V voltage tree vector (of order b) which elements include the known branch voltages (input data: v-branches, i, v-branches, and v, f-branches),
- V_f voltage tree vector which elements can be found using known v-tree impedances taken from f-branches and f, i-branches and using the unknown voltage state vector $I_{m,o}$.

$$V_f = Z_n T_i I_{m,o} \tag{13.6}$$

where Z_n—impedance matrix of v-tree branches.

Substituting Eqs. (13.4) into (13.3) and (13.6) into (13.5), we have

$$I_{m,o} = I_i + Y_m T_v V_{o,n} \tag{13.7}$$

$$V_{o,n} = V_v + Z_n T_I I_{mo} \tag{13.8}$$

Note that in the current state vector $I_{m,o}$ and the voltage state vector $V_{o,n}$, both are in each of the above equations, so there are possible two substitutions. Substituting (13.8) into (13.7) and substituting (13.7) into (13.8), we have

$$I_{m,o} = I_i + Y_m T_v (V_v + Z_n T_i I_{m,o}) \tag{13.9}$$

$$V_{o,n} = V_i + Z_n T_i (I_i + Y_m T_v V_{o,n}) \tag{13.10}$$

Solving Eq. (13.9) for $I_{m,o}$ and solving Eq. (13.10) for $V_{o,n}$, the current state vector and the voltage state vectors can be found as follows:

$$I_{m,o} = \left(1^b - Y_m T_v Z_n T_i\right)^{-1}(I_i + Y_m T_v V_v) \tag{13.11}$$

$$V_{o,n} = \left(1^b - Z_n T_i Y_m T_v\right)^{-1}(V_v + Z_n T_i I_i) \tag{13.12}$$

Denoting:

$$F_y = \left(1^b - Y_m T_v Z_n T_i\right) \tag{13.13}$$

$$F_z = \left(1^b - Z_n T_i Y_m T_v\right) \tag{13.14}$$

The final current state vector and voltage state vector are as follows:

$$I_{m,o} = F_y^{-1}(I_i + Y_m T_v V_v) \tag{13.15}$$

$$V_{o,n} = F_z^{-1}(V_v + Z_n T_i I_i) \tag{13.16}$$

13.3 Solvability Analysis

Let us now consider the solvability conditions. In the above equations, vectors of the current and voltage states $I_{m,o}$ and $V_{o,n}$ are to be found. Vectors I_i, V_v and matrices Y_c Z_t are formed from the given input data.

In order to find the current state vector, the number of nonzero entries in the current cotree vector $I_{m,o}$ must be equal to *m*. In other words, the current topological structure, represented by topological matrix T_i, must be chosen in such a way that the *i*-cotree include *i*-branches, *i, v*-branches, and *i, f*-branches and if necessary *f*-branches and *f, v*-branches, but can not include *o*-branches. The branches with known current and admittance (*i, f*-branches) must be treated as branches with known current, and *f, v*-branches must be treated as branches with known admittance only.

In order to find the voltage state vector, the number of nonzero entries in the voltage tree vector $V_{o,n}$ must be equal to *n*. In other words, the voltage topological structure, represented by topological matrix T_v, must be chosen in such a way that the *v*-tree include *v*-branches, *i, v*-branches, and *v, f*-branches and if necessary *f, i*-branches and *f*-branches, but can not include *o*-branches. The branches with known voltage and impedance (*v, f*-branches) must be treated as branches with known voltage only and *i, f*-branches as known impedance only.

The additional necessary solvability condition is the non-singularity of matrices F_y and F_z. Note that if the branch *j*, with given parameter (admittance/impedance), belongs to the current *i*-cotree and also to the voltage *v*-tree (current topology and voltage topology are non-complementary), what means that the entry z_j in matrix Z_t is in the same time the entry $y_j = 1/z_j$ in matrix Y_c, then as a result of multiplication (taking into account matrices T_v and T_v) the diagonal entries j of matrices F_z and F_y are equal to zero. It may mean singularity of matrices. It leads to the following solvability condition: If branch parameter is used as impedance in matrix F_z, then it can not be used as admittance in matrix F_y and opposite. If such condition is fulfilled, then the diagonal entries of matrices F_z and F_y are equal to 1 and

$$F_y^{-1} I_i = I_i \tag{13.17}$$

$$F_z^{-1} V_v = V_v \tag{13.18}$$

The final current state and voltage state equations are as follows

$$I_{m,o} = I_i + F_y^{-1} Y_m T_v V_v \tag{13.19}$$

$$V_{o,n} = V_v + F_z^{-1} Z T_i I_i \tag{13.20}$$

and the final current and voltage equations of a network are as follows

$$I = T_i \left(I_i + F_y^{-1} Y_m T_v V_v \right) \tag{13.21}$$

$$V = T_v \left(V_v + F_z^{-1} Z_n T_i I_i \right) \tag{13.22}$$

In practical applications, the independent solution of currents and voltages is not necessary. The vector I can be found from Eq. (13.21) and vector V from simple equation avoiding calculation of matrix F_z^{-1} as follows

$$V = T_v (V_v + Z_n I) \tag{13.23}$$

or opposite; if vector V is found from Eq. (13.22), then vector I (avoiding calculation of matrix F_y^{-1}) can be found from simple formula

$$I = T_i (I_i + Y_m V) \tag{13.24}$$

13.4 Numerical Example

The input data of the arbitrary method of network solution are illustrated on a simple example of network shown in Fig. 13.1. There is given network graph and following branch data: one *i, v*-branch, one *i*-branch, one *v*-branch, one *f, i*-branch, and one *v, f*-branch; for three branches, there are not any input data.

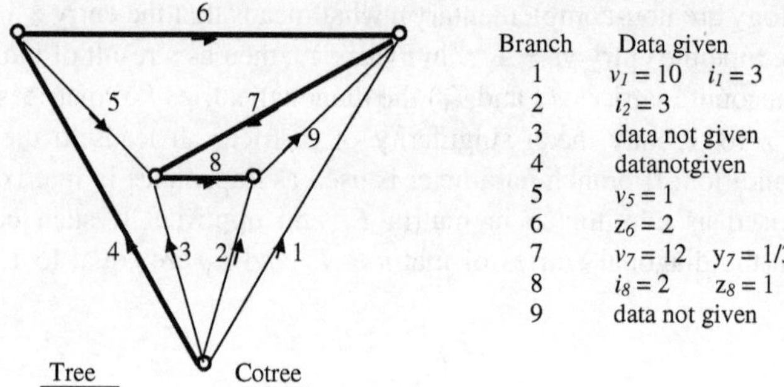

Fig. 13.1 Input data of arbitrary method of network solution

As it follows from the solvability conditions, there are two topological structures necessary. The tree of voltage topology must include branches 1 and 7 because of given voltages v_1 and v_7. The cotree of current topology must include branches 1, 2, and 8 because of given currents i_1, i_2, and i_8. The branches with not given data can not belong to voltage tree and current cotree. The voltage topology and the current topology are shown in Fig. 13.2.

The numerical calculations must be done using only one topology of network. Any one is possible but one of two in Fig. 13.2 is recommended. Assuming current tree/cotree topology, the ordering of branches in vectors and matrices is as follows: First, $m = 5$ branches are cotree branches numbered 1, 2, 5, 7, and 8, and the last $n = 4$ branches are tree branches numbered 3, 4, 6, and 9. The matrices T_I, T_V, Y_m, Z_n (see Eqs. 13.13 and 13.14) are given below. Note that the ordering of branches in matrices must be the same

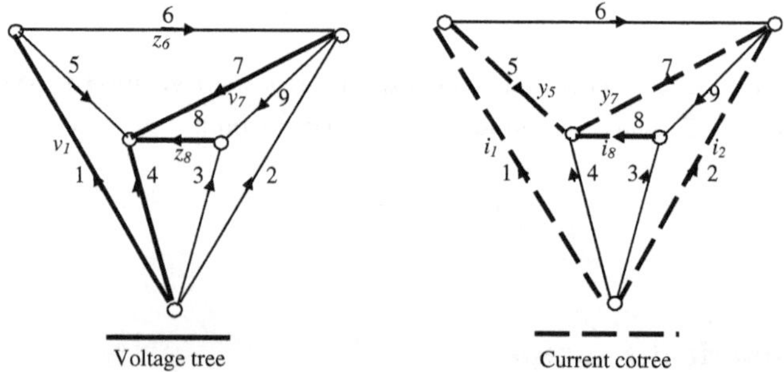

Fig. 13.2 Chosen voltage and current topologies

$$
\begin{array}{r@{\;}c}
& \begin{array}{ccccccccc} 1 & 2 & 5 & 7 & 8 & 3 & 4 & 6 & 9 \end{array} \\
T_I = & \left[\begin{array}{ccccccccc}
1 & 0 & 0 & 0 & 0 & 1 & 0 & -1 & -1 \\
0 & 1 & 0 & 0 & 0 & 1 & 0 & 0 & -1 \\
0 & 0 & 1 & 0 & 0 & -1 & 1 & 1 & 1 \\
0 & 0 & 0 & 1 & 0 & -1 & 1 & 0 & 1 \\
0 & 0 & 0 & 0 & 1 & -1 & 1 & 0 & 0 \\
-1 & -1 & 1 & 1 & 1 & 1 & 0 & 0 & 0 \\
0 & 0 & -1 & -1 & -1 & 0 & 1 & 0 & 0 \\
1 & 0 & -1 & 0 & 0 & 0 & 0 & 1 & 0 \\
1 & 1 & -1 & -1 & -1 & 0 & 0 & 0 & 1
\end{array}\right]\begin{array}{c} 1 \\ 2 \\ 5 \\ 7 \\ 8 \\ 3 \\ 4 \\ 6 \\ 9 \end{array}
\end{array}
$$

$$
\begin{array}{r@{\;}c}
& \begin{array}{ccccccccc} 1 & 2 & 5 & 7 & 8 & 3 & 4 & 6 & 9 \end{array} \\
T_V = * & \left[\begin{array}{ccccccccc}
1 & 0 & 1 & 0 & 0 & 0 & 0 & 1 & 0 \\
0 & 1 & 0 & -1 & 0 & 0 & 1 & 0 & -0 \\
-1 & 0 & 1 & 0 & 0 & 0 & -1 & 0 & 0 \\
0 & 1 & 0 & 1 & 0 & 0 & 0 & 1 & -0 \\
0 & 0 & 0 & 0 & 1 & 1 & 0 & 0 & 1 \\
0 & 0 & 0 & 0 & -1 & 1 & 1 & 0 & 0 \\
0 & -1 & 1 & 0 & 0 & -1 & 1 & 1 & 0 \\
-1 & 0 & 0 & -1 & 0 & 0 & -1 & 1 & 0 \\
0 & 0 & 0 & 1 & -1 & 0 & 0 & 0 & 1
\end{array}\right]\begin{array}{c} 1 \\ 2 \\ 5 \\ 7 \\ 8 \\ 3 \\ 4 \\ 6 \\ 9 \end{array}
\end{array}
$$

$$
\begin{array}{r@{\;}c}
& \begin{array}{ccccccccc} 1 & 2 & 5 & 7 & 8 & 3 & 4 & 6 & 9 \end{array} \\
Y_m = & \left[\begin{array}{ccccccccc}
0 & 0 & 0 & 0 & 0 & 0 & 0 & 0 & 0 \\
0 & 0 & 0 & 0 & 0 & 0 & 0 & 0 & -0 \\
0 & 0 & \boldsymbol{1} & 0 & 0 & 0 & 0 & 0 & 0 \\
0 & 0 & 0 & \boldsymbol{1/2} & 0 & 0 & 0 & 0 & 0 \\
0 & 0 & 0 & 0 & 0 & -0 & 0 & 0 & 0 \\
0 & 0 & 0 & 0 & 0 & 0 & 0 & 0 & 0 \\
0 & 0 & 0 & 0 & 0 & 0 & 0 & 0 & 0 \\
0 & 0 & 0 & 0 & 0 & 0 & 0 & 0 & 0 \\
0 & 0 & 0 & 0 & 0 & 0 & 0 & 0 & 0
\end{array}\right]\begin{array}{c} 1 \\ 2 \\ 5 \\ 7 \\ 8 \\ 3 \\ 4 \\ 6 \\ 9 \end{array}
\end{array}
$$

$$
Z_n = \begin{array}{c} \\ \left[\begin{array}{ccccccccc} 0 & 0 & 0 & 0 & 0 & 0 & 0 & 0 & 0 \\ 0 & 0 & 0 & 0 & 0 & 0 & 0 & 0 & -0 \\ 0 & 0 & 0 & 0 & 0 & 0 & 0 & 0 & 0 \\ 0 & 0 & 0 & 0 & 0 & 0 & 0 & 0 & 0 \\ 0 & 0 & 0 & \mathbf{1} & -0 & 0 & 0 & 0 & 0 \\ 0 & 0 & 0 & 0 & 0 & 0 & 0 & 0 & 0 \\ 0 & 0 & -0 & 0 & 0 & 0 & 0 & 0 & 0 \\ 0 & 0 & 0 & 0 & 0 & 0 & 0 & \mathbf{2} & 0 \\ 0 & 0 & 0 & 0 & 0 & 0 & 0 & 0 & 0 \end{array}\right] \end{array}
\begin{array}{c} 1 \\ 2 \\ 5 \\ 7 \\ 8 \\ 3 \\ 4 \\ 6 \\ 9 \end{array}
$$

(columns: 1 2 5 7 8 3 4 6 9)

The voltage and current input data are as follows:

$$
\begin{array}{cccccccccc} & 1 & 2 & 5 & 7 & 8 & 3 & 4 & 6 & 9 \\ I_i = [& 10 & 0 & 0 & -2 & 0 & 0 & 0 & 0 & 0] \end{array}
$$

$$
\begin{array}{cccccccccc} & 1 & 2 & 5 & 7 & 8 & 3 & 4 & 6 & 9 \\ V_V = [& -6 & 1 & 0 & 0 & 0 & 0 & 0 & 0 & 0] \end{array}
$$

After substituting the input data into the solution equations

$$
I = T_i(I_i + F_y^{-1} Y_m T_v V_v)
$$

$$
V = T_v(V_v + F_z^{-1} Z_n T_i I_i)
$$

The final results in the form of current and voltage vectors are as follows:

$$
\begin{array}{cccccccccc} & 1 & 2 & 5 & 7 & 8 & 3 & 4 & 6 & 9 \\ I = [& -6 & 1 & 1 & -1 & 2 & 2 & 3 & 2 & 3] \end{array}
$$

$$
\begin{array}{cccccccccc} & 1 & 2 & 5 & 7 & 8 & 3 & 4 & 6 & 9 \\ V = [& 10 & 9 & 1 & -2 & 1 & 8 & 7 & 3 & 1] \end{array}
$$

Note that all unknown branch parameters (impedances/admittances) can be easily found from the above results.

The main part of computer program is the procedure of finding the needed current and voltage topologies. Algorithm of finding simple tree and complementary cotree of a graph is not a problem. Finding the cotree separate, not as a complement to the tree, generally is very difficult (it is possible in special graphs only). Finding the current cotree and the voltage tree, according to the solvability conditions, can be done using various algorithms. Note that both topologies can be found using the procedure of finding a tree of graph only (voltage tree from *v*-branches, *i, v*-branches, *v, f*-branches, and *f*-branches and a current cotree which is a complement to a tree from *v*-branches, *v, f*-branches, *o*-branches, and *f*-branches).

Generally, the arbitrary input data method can be applied to the solution of problems with various distributions of input data. For example, in power system operation problems, the method can be used in the problem of power system state estimation.

Chapter 14
Logical Optimization Method of Network System

14.1 Example of Network System Matrices

In this chapter, the logical method of finding the quasi-optimal state of network system is derived, using the system admittance and system impedance matrices Y_S and Z_S of a network [see Chap. 8, Eqs. (8.37) and (8.38)]. The logical method is illustrated on a simple example of nine-branch power system network. The input data are the system admittance Y_S and system impedance Z_S, which are found using the relations derived in Sect. 8.3 as follows:

1. The starting point is the chosen topology of network system and the current and voltage vectors I and V in Fig. 14.1.
   The topology of network may have essential influence on the results of computations. The star-like form of tree used in power system networks (tree branches are node-to-earth branches) may lead to negative results. In example of network in Fig. 14.1, we assume that the tree consists of branches *4*, *6*, *7*, and *8* and the cotree consists of branches *1*, *2*, *3*, *5*, and *9*.
2. The active and passive branch parameters should be chosen, particularly the ideal current sources in cotree and ideal voltage sources in tree. The remaining branches are impedances in cotree and admittances in tree. In our example, we assume that currents of branches *1* and *2* are ideal current sources and voltage of branch *4* is an ideal voltage source.
3. Using the elements of current vector I and elements of voltage vector V (see Fig. 14.1), the values of branch impedances and branch admittances are calculated. It makes possible the formulation of following network parameters: matrix $F_{m,n}$ [see Chap. 8, Eq. (8.24)] and network system state vector $K_{I,V,F}$ [see Chap. 8, Eq. (8.24)]. The ideal sources in matrix $F_{m,n}$ are indicated by number 1, and numerical values of sources are elements of vector $K_{I,V,F}$.
4. Using inverted topological matrix T^{-1}, the equation $F_{m,n}T^{-1}I + F_{m,n}T^{-1}V = K_{m,n}$ can be formulated [see Eqs. (8.25)–(8.29)].

A. Kłos, *Mathematical Models of Electrical Network Systems*, Lecture Notes in Electrical Engineering 412, DOI 10.1007/978-3-319-52178-7_14

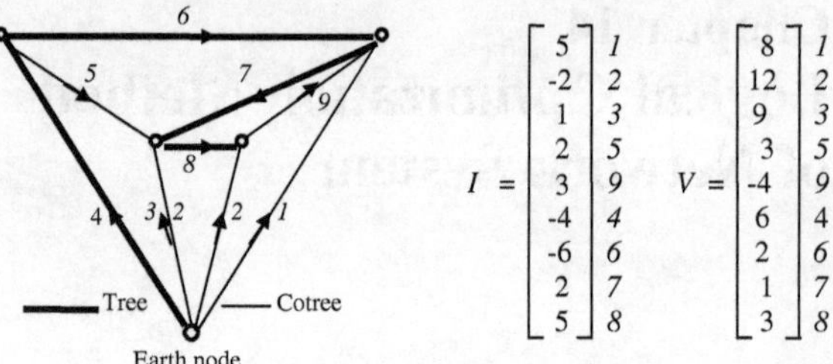

Fig. 14.1 Network graph and current voltage state of network

5. Reducing vector $K_{m,n}$ to zero by transferring currents and voltages from this vector to the vectors I and V, we have $R_I I + R_V V = 0$ [see Eqs. (8.30) and (8.31)].
6. Finally, the system admittance matrix Y_S and the system impedance matrix Z_S are found from equations $Y_S = -R_I^{-1} R_V$, $Z_S = -R_V^{-1} R_I$, and the system equation $I = Y_S V$ and $V = Z_S I$ are as follows:

$$
\begin{bmatrix} 5 \\ -2 \\ 1 \\ 2 \\ 3 \\ -4 \\ -6 \\ 2 \\ 5 \end{bmatrix}
=
\begin{bmatrix}
\overset{1}{-4.000} & \overset{2}{-0.167} & \overset{3}{0.111} & \overset{5}{0.667} & \overset{9}{-0.000} & \overset{4}{4.000} & \overset{6}{1.000} & \overset{7}{-0.000} & \overset{8}{0.000} \\
-1.000 & 0.750 & 0.111 & 0.667 & 1.875 & 0.000 & 0.000 & 1.000 & 1.500 \\
-1.000 & -0.167 & 1.000 & 0.667 & 0.750 & 0.000 & 0.000 & 1.000 & 0.000 \\
-1.000 & -0.167 & 0.111 & 1.000 & 0.750 & 0.000 & 0.000 & 1.000 & 0.000 \\
0.000 & 0.147 & -0.111 & -0.667 & -0.125 & 0.000 & 0.000 & -1.000 & -1.500 \\
4.000 & 0.000 & 0.000 & -0.667 & 0.000 & -4.000 & -1.000 & 0.000 & -0.000 \\
3.000 & 0.000 & 0.000 & 0.000 & 0.000 & -3.000 & -0.000 & -0.000 & 0.000 \\
-0.000 & -0.333 & 0.222 & 1.133 & 1.500 & 0.000 & 0.000 & 2.000 & 0.000 \\
-0.000 & -0.147 & 0.000 & -0.000 & 1.875 & 0.000 & 0.000 & -0.000 & 2.500
\end{bmatrix}
\begin{bmatrix} 8 \\ 12 \\ 9 \\ 3 \\ -4 \\ 6 \\ 2 \\ 1 \\ 3 \end{bmatrix}
\begin{matrix} 1 \\ 2 \\ 3 \\ 5 \\ 9 \\ 4 \\ 6 \\ 7 \\ 8 \end{matrix}
$$

$$I = \qquad Y_S \qquad V$$

$$
\begin{bmatrix} 8 \\ 12 \\ 9 \\ 3 \\ -4 \\ 6 \\ 2 \\ 1 \\ 3 \end{bmatrix}
=
\begin{bmatrix}
\overset{1}{-56.000} & \overset{2}{-8.000} & \overset{3}{7.000} & \overset{4}{0.000} & \overset{5}{0.000} & \overset{6}{-56.000} & \overset{7}{-0.000} & \overset{8}{0.500} & \overset{9}{4.800} \\
-48.000 & -6.000 & 6.000 & 0.000 & 0.000 & -48.000 & 0.000 & 0.000 & 3.600 \\
-63.000 & -9.000 & 9.000 & 0.000 & 0.000 & -63.000 & 0.000 & 0.000 & 5.400 \\
-0.000 & 0.000 & 0.000 & 3.000 & 0.000 & -0.000 & 1.000 & -1.500 & 0.000 \\
-0.000 & 0.000 & -0.000 & 0.000 & 0.571 & -0.000 & -0.000 & -0.286 & 0.343 \\
-56.000 & -8.000 & 7.000 & 0.000 & 0.000 & -56.000 & -0.333 & 0.500 & 4.000 \\
0.000 & 0.000 & -0.000 & -2.000 & 0.000 & -1.000 & 0.667 & 1.000 & -0.000 \\
-1.000 & 0.000 & -0.000 & -2.000 & -0.429 & -1.000 & 0.667 & 1.286 & 0.257 \\
-8.000 & -1.000 & 1.000 & 0.000 & -0.429 & -8.000 & -0.000 & -1.214 & 0.743
\end{bmatrix}
\begin{bmatrix} 5 \\ -2 \\ 1 \\ 2 \\ 3 \\ -4 \\ -6 \\ 2 \\ 5 \end{bmatrix}
\begin{matrix} 1 \\ 2 \\ 3 \\ 5 \\ 9 \\ 4 \\ 6 \\ 7 \\ 8 \end{matrix}
$$

$$I = \qquad Y_S \qquad V$$

Equations $I = Y_S V$ and $V = Z_S I$ make possible the analytical investigation of current voltage functions in a network system. Elements of matrix Y_S are *system self and mutual admittances* of network branches, and elements of matrix Z_S are *system self and mutual impedances* of network ranches. They are the coefficients of branch currents as a function of branch voltages and coefficients of branch voltages as a function of branch currents. Mathematically, they are derivatives *di*/*dv* and *dv*/*di* of current voltage functions. Generally, the elements of matrices Y_S and Z_S characterize the quantitative influence of voltages on currents and currents on voltages in the network system.

14.2 Logical Analysis of Network System Equations

In this section, the analysis of system equations is illustrated on a simple example of the state of network as shown in Fig. 14.1. Various analytical methods may be used, but it is of advantage using numerical method of how much currents change as a function of unit voltage values and of how much voltages change as a function of unit current values. It means that equations $I = Y_S V$ and $V = Z_S I$ should be substituted by suitable coefficients. Consider equation $I = Y_S V$. Each element $y_{j,k}$ of matrix Y_S is a coefficient of equation:

$$i_{j,k} = y_{j,k} v_k$$

Current $i_{j,k}$ is a part of current i_j of branch j, and v_k is an element k of voltage vector V. Reducing this dependence $i_{j,k}$ on v_k to the unit value of voltage v_k, we have *the current-per-unit-voltage coefficient* $s^i_{j,k}$.

$$s^i_{j,k} = y_{j,k}/v_k$$

Analogically, in equation $V = Z_S I$, each element $z_{j,k}$ of matrix Y_S is a coefficient of equation:

$$v_{j,k} = z_{j,k} i_k$$

Voltage $v_{j,k}$ is a part of voltage v_j of branch j and i_k is an element k of current vector I. Reducing this dependence $v_{j,k}$ on i_k to the unit value of current i_k, we have *the voltage-per-unit-current coefficient* $s^v_{j,k}$.

$$s^v_{j,k} = z_{j,k}/i_k$$

Using coefficients $s^i_{j,k}$ $s^v_{j,k}$, one can solve various practical problems. Let us analyze current voltage interrelations in some branches of a network system in Fig. 14.1.

1. The first row of equation $I = Y_S V$ concerns current of branch number *1* as a function of voltage vector V. Branch *1* is an ideal current source; if it is not a part of network, then it produces constant current and its admittance is equal to zero. However, branch *1*, if it is an ideal current source in the network system, has system self and system mutual admittances, which are "thrown upon" on branch *1* by a whole network system. It means that in network system, the branch number *1* is as well the ideal current source as the admittance branch. The nominal value of the ideal current source of branch *1* remains unchanged, but it is a function of network voltages (elements of vector V) and the system self and system mutual admittances. Numerically, the analysis of how the

current of branch *1* depend on network voltages can be done using the coefficients $s^i_{1,k}$ which are as follows:

Branches	*1*	2	*3*	*5*	9	*4*	6	7	*8*
$s^i_{1,k}$	−0.500	−0.014	0.014	0.222	0.000	0.667	0.500	0.000	0.000

It means that the greater positive influence on the current of branch *1* has the voltage of branch *4*, and the greater negative influence has the voltage of branch *1*. Not any influence has voltages of branches *9*, *7*, and *8*. If the current of branch *1* is to large then in order to decrease this current it is of advantage to decrease the value of ideal voltage source of branch *4* and decrease voltage of branch *6*, and if it is not possible then increase voltage of branch *1* and decrease voltage of branch *6*. The qualitative estimate of how much voltage value should be increase can be done using coefficients $s^i_{1,k}$.

2. The first row of equation $V = Z_S I$ concerns voltage of branch *1* as a function of voltage vector *V*. Branch *1*, if it is an ideal current source in the network system, has system self and system mutual impedances, so branch *1* is as well the ideal voltage source as the impedance branch. The nominal value of ideal current source of branch *1* is an element of current vector *I*. The voltage of branch *1* is a function of network currents (elements of vector *I*) and the system self and system mutual impedances of branch *1*. Numerically, the analysis of how the voltage of branch *1* depend on network current vector *1* can be done using the coefficients $s^v_{1,k}$. Such coefficients for the first row of equation $V = Z_S I$ are as follows:

Branches	*1*	2	*3*	*5*	9	*4*	6	7	*8*
$s^v_{1,k}$	−11.200	4.000	7.000	0.000	0.000	14.000	0.000	0.250	0.960

It means that the greater positive influence over the voltage of branch number *1* has the current of branch *4*, and the greater negative influence has the current of branch *1*. Not any influence has currents of branches *5*, *9*, and *6*. If the power of branch number *1* must be increase, taking into account the coefficients $s^i_{1,k}$ and $s^v_{1,k}$, then the best operation is to decrease the value of ideal voltage source of branch *4* and to decrease the current of branch *4*, and if it is not possible then decrease the current of branch *3* and increase the current of branch *1*. The qualitative estimate of how much power value should be increase can be done using coefficients $s^i_{1,k}$ and $s^v_{1,k}$.

3. The third row of equation $I = Y_S V$ concerns the current of branch *3* as a function of voltage vector *V*. Branch *3*, if it is not a part of network, has its own self-admittance equal to 1/9 (no mutual admittances), but if it is a part of network in Fig. 14.1 then the system self-admittance of branch *3* has also the

same value 1/9. It is not a rule, because the same kind branch number *5* has the self-admittance equal to the system self-admittance. Additionally, branch *3* has system mutual admittances, which characterize the difference between the classical self and mutual branch parameters and the system self and system mutual branch parameters. The current of branch *3* is a sum of products of network voltages (vector V) and system self and system mutual admittances of branch *3*. Numerically, the analysis of how the current of branch *3* depend on network voltage vector V can be done using the current-per-unit-voltage coefficients $s^i_{3,k}$ which are as follows:

Branches	*1*	2	*3*	*5*	*9*	*4*	*6*	*7*	*8*
$s^i_{3,k}$	−0.125	−0.014	0.111	0.222	−0.188	0.000	0.000	1.000	0.000

The greater positive influence over the current of branch *3* has the voltage of branch *7*, and not any influence has the voltages of branches *4*, *6*, and *8*. In case if current of branch *3* is too small, then the optimal operation is to increase the voltage of branch *7*. Note that the ideal voltage source of branch *4* cannot be used because $s^i_{3,4} = 0$.

4. The third row of equation $V = Z_S I$ concerns the voltage of branch *3* as a function of current vector *I*. Branch *3*, if it is not a part of network, has its own self-impedance equal to 9 (no mutual impedances), and if it is a part of network then its system self-impedance has the same value, but additionally branch *3* has system mutual impedances. The current of branch *3* is a sum of products of network currents and system self and system mutual impedances. Numerically, the analysis of how the voltage of branch *3* depends on network currents can be done using the voltage-per-unit-current coefficients $s^v_{3,k}$.

Branch	*1*	2	*3*	*5*	*9*	*4*	*6*	*7*	*8*
$s^v_{3,k}$	−12.600	4.500	0.9.000	0.000	0.000	15.750	0.000	0.000	1.080

It means that the greater positive influence over the voltage of branch *3* has the current of branch *4*, and not any influence has voltages of branches *5*, *9*, *6*, and *7*.

If the power of branch *3* should be increase, instead of diminishing current of branch *3* and voltage of branch *3*, then it is better to decrease the current of branch *4* and decrease the voltage of branch *7* (see coefficients $s^v_{3,k}$ and $s^i_{3,k}$).

5. The sixth row of equation $I = Y_S V$ concerns the current of branch *4* (which is ideal voltage source) as a function of voltage vector *V*. If branch *4* is not a part of network, then it is constant ideal voltage, and its admittance is equal to zero. Branch *4*, as an ideal voltage source in the network system, has system self and system mutual admittances, which are "thrown upon" on branch by a whole network system, so it is both voltage source and admittance branch. The

nominal value of an ideal voltage source is an element of voltage vector *V*. Numerically, the analysis of how the current of branch *4* depends on network voltages can be done using the $s^i_{4,k}$.

Branch	*1*	2	3	5	9	4	6	7	8
$s^i_{4,k}$	−0.500	−0.000	0.000	0.222	0.000	−0.667	−0.500	0.000	0.000

It means that the greater negative influence on the current of branch *4* has the ideal voltage source of branch *4*, and the greater positive influence has the voltage of branch *5*. Not any influence has voltages of branches *2*, *3*, *9*, *7*, and *8*. In case if the current of branch *4* should be increase, then the best operation is increasing the nominal value of ideal voltage source of branch *4*: If it is not possible, then voltages of branch *1* and branch *6* should be increased.

6. The sixth row of equation $V = Z_S I$ concerns voltage of branch *4* (which is ideal voltage source) as a function of current vector *I*. Branch *4*, if it is not a part of network, produces constant voltage and its impedance is equal to zero. Branch *4*, as a part of network system, has system self and system mutual impedances, which are thrown upon on branch by whole network system, so it is both the ideal voltage source and system impedance branch. The constant value of ideal voltage source of branch *4* is an element of voltage vector *V* and is a sum of products of network currents and system self and system mutual impedances. The analysis of how the voltages of branch *4* depend on network currents can be done using $s^v_{j,k}$:

Branch	*1*	2	3	5	9	4	6	7	8
$s^v_{4,k}$	−11.200	4.000	7.000	0.000	0.000	14.000	0.000	0.250	0.960

The greater negative influence on the voltage of branch *4* has the current of branch *4*, and the greater negative influence has the current of branch *1*. Not any influence has currents of branches *5*, *9*, and *6*.

If the power of branch *3* should be increase, then the current of branch *4* should be increase and the ideal voltage source of branch *4* should be increased.

Note, e.g., that the coefficient $s^v_{4,1} = -11.200$ has much higher negative value then the coefficient $s^i_{4,1} = 0.5$, and the coefficient $s^v_{4,4} = 14.000$ has much higher positive value then the coefficient $s^i_{4,4} = 0.667$. Generally, it means that the currents have greater influence on voltages and voltages have littler influence on currents.

7. The row number *9* of equation and $V = Z_S I$ concerns the voltage of branch *8* as a function of current vector *I*. Branch *8*, if is not a not a part of network, has its own self-impedance and no mutual impedances. If it is a part of our network then has system self-impedance and system mutual impedances thrown upon by

a whole network system. Numerically, the analysis of how the voltage of branch *8* depends on network currents can be done using the $s^{v}_{8,k}$ coefficients.

Branch	*1*	2	*3*	*5*	9	*4*	*6*	*7*	*8*
$s^{v}_{8,k}$	−1.600	−0.500	1.000	0.000	−0.143	2.000	0000	−0.107	0.143

It means that the greater positive influence on the voltage of branch *8* has the current of branch *4*; and the greater negative influence has the current of branch *1*. Not any influence has the voltages of branches *5* and *6*. In case if the voltage of branch *4* should be increase, then the best operation is diminishing the current of branch *4* and increasing the nominal value of ideal current source of branch *1*. If it is not possible, then currents of branch *1* should be increased and of branch *3* should be increase.

8. The row number *9* of equation $I = Y_S V$ concerns current of branch *8* as a function of voltage vector *V*. Branch *8*, if it is a part of our network, has system self-admittance and system mutual admittances thrown upon on branch *8* by whole network system. The current of branch *8* is a sum of products of network voltages and system self and system mutual impedances. The analysis of how the current of branch *8* depends on network voltages can be done using the $s^{i}_{8,k}$ coefficients which are as follows:

Branch	*1*	2	*3*	*5*	9	*4*	6	*7*	*8*
$s^{i}_{8,k}$	0.000	0.035	0.000	0.000	−0.469	0.000	0000	0.000	0.833

It means that branch *8* has system self-admittance and only two system mutual admittances. The greater positive influence on the current of branch *8* has the voltage of branch *8*, and the greater negative influence has the voltage of branch *9*. If the power of branch *8* should be increase, then the current of branch *4* and the voltage of branch *8* should be increase.

Each of the above-presented examples of logical analysis and application equations $I = Y_S V$ and $V = Z_S I$ is the first step only in the logical method of finding the quasi-optimal state of network system. Note that if in any of the above examples the branch current or branch voltage is changed, then current and voltages in all branches are changed. If the vectors and matrices, before and after change, are not in prescribed limits, then calculation must be repeated. The new current vectors *I* and *V*, and new matrices Y_S and Z_S are to be found, and the logical analysis as above should be repeated in iteration process.

14.3 Application Possibilities

The applications of a method illustrated in previous section are restricted, because in some cases the matrices Y_S and Z_S are singular, and the equations $I = Y_S V$, $V = Z_S I$ does not exist. It depends mainly on the number and configuration of ideal current and voltage sources in a network system. In order to illustrate the application possibilities, the various cases of number and distribution of ideal sources are done using a simple example of network system as shown in Fig. 14.1. In Sects. 8.1 and 8.2, it was shown that using the input data and assuming the ideal sources in branches in branches *1*, *2*, and *4*, the matrices Y_S and Z_S are non-singular, and the logical analysis is possible. The same results are obtained using the same input data and assuming up to three ideal sources located differently in network branches. Following computations were done.

- 1 source located in branch *3* or in branch *6* or in branch *8*.
- 2 sources located in branches *3*, *5*, in branches *2*, *6*, or in branches *5*, *8*.
- 3 sources located in branches *1*, *2*, *3* or in branches *4*, *6*, *9*.

The same (positive) results are obtained if 4 sources are assumed located in branches *1*, *2*, *3*, *5* or branches *4*, *6*, *7*, *8* only, but for other locations of 4 sources results are negative.

If 4 sources are located in branches *1*, *2*, *3*, *6*, then matrix Y_S is non-singular and matrix is singular. It means that equation $I = Y_S V$ exists and equation $V = Z_S I$ does not. The logical analysis is restricted.

If 4 sources are located in branches *4*, *6*, *7*, *8*, then matrix Z_S is non-singular and matrix $Y_S Y_S$ is singular. It means that equation $V = Z_S I$ exists and equation $I = Y_S V$ does not. The logical analysis is restricted.

If 5 and more sources are assumed, then the relation between the equations $I = Y_S V$ and $V = Z_S I$ does not exist and the logical analysis cannot be done.

Interesting is the case of network system in Fig. 14.1, in which there are no sources and all branches are current–voltage functions only. In this case, both matrices Y_S and Z_S are non-singular, and the equations $I = Y_S V$ and $V = Z_S I$ does exist, so the logical analysis can be done. Note the important difference between the classical matrices *Y*, Z, and the system matrices Y_S and Z_S in this case. The elements of classical matrices are pure branch parameters independent of network, and the elements of system matrices are system parameters dependent of network system. In our case, the elements of the first one are the self parameters of branches, independent of network, and the elements of the last one are the self and mutual system parameters, which are “thrown upon” on the elements by network system. It illustrates the difference between a network as a sum of branches and a network system. Network system is a unique physical quantity, which has its own features and laws.

As it comes out from the above, the logical method of finding the quasi-optimal state of network system can be computerized and practically implemented in electrical power systems operation and development. Such method may help, e.g., to assess the reliability of power system in the case of over and under-loads, or over and under-voltages of network branches (lines, transformers). The power system operator, having the current and voltage state of power system (e.g., from the state estimation) and using computerized method presented above, can be nearly immediately provided with necessary information. Such a tool may help to restore the power system security in cases of faults including blackouts.

Chapter 15
Current-Based Method of Load Flow Solution

Using classical input data, the process of finding the load flow in power system networks is done in two steps. First, the voltage state vector is calculated; second, the needed output quantities (branch power loads) are found. In recently used methods, the voltage state vector, which is a vector of all nodal (node-to-earth) voltages, is calculated using iteration process. However, taking into account the algebraic model of network system described in Part I of book and the general network solution method derived in Part II Chap. 11, the load flow can be found using current state vector. The current-based load flow method for any electrical network can be derived using general network solution method (Sect. 11.1). The derivation may be complicated because the b-dimensional Eqs. (11.15) and (11.16) must be reduced to n and m dimensional. In case of power system load flow method, the input data are known (classical input data), so it simplifies the derivation of method. However, in order to find the current state vector, the suitable, current state-oriented topology of network (different than the classical one) should be chosen. In this chapter, *the current-based load flow method (CBLF)* is derived and verified.

15.1 Current State-Oriented Topology of Network

Consider any power system network. Suppose that there are given the classical load flow input data. If there are known load values of all node-to-earth branches $s_{n,j}$ and nominal network voltage V_n, then the nodal current values $i_{n,j} = s_{n,j}/V_n$ are very good estimate of real current values. Such nodal currents vector could not be used as the current state vector because it is not a current vector of any set of cotree branches. However, such vector can help to find the current state-oriented topology of network. In order to use the nodal current vector as a part of current state vector, the elements of nodal current vector must belong to the currents of cotree branches.

A. Kłos, *Mathematical Models of Electrical Network Systems*,
Lecture Notes in Electrical Engineering 412,
DOI 10.1007/978-3-319-52178-7_15

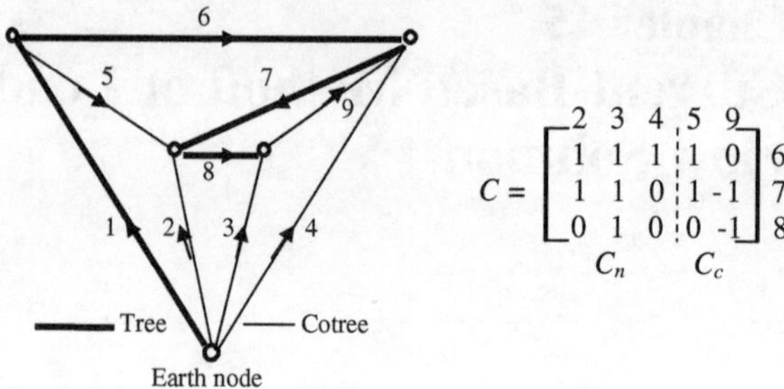

Fig. 15.1 Illustration of current state-oriented topology of network

Let us subdivide a network graph into a tree and a cotree as it is shown in Fig. 15.1. The set of tree branches include one node-to-earth branch [a slack node branch (in Fig. 15.1 branch *1*)] and d freely chosen node-to-node branches (in Fig. 15.1 branches *6*, *7*, *8*). The cotree include the remaining branches, among them: n node-to-earth branches (excluding slack-node-to-earth branch), and c node-to-node branches which doesn't belong to a tree (in Fig. 15.1 branches *2*, *3*, *4*, *5*, *9*).

A tree defined as above determines d independent cut-sets of a network, and each of them is defined by a tree branch. Each network cut-set includes one tree branch and a set of cotree branches parting this cut-set from the rest of network. The cut-set incidence matrix C, of order $d \times (n + c)$, relates cotree branches (columns) to tree branches (rows). Each row of C represents the cut-set defined by a tree branch. Note that the submatrix C_m transposed is a loop-set incidence matrix in the classical node-to-node network. Matrix C can be written as follows:

$$C = [\,C_n \quad C_m\,] \tag{15.1}$$

where

C_n submatrix relating tree branches to node-to-earth cotree branches of order $d \times n$ and

C_c submatrix relating tree branches to node-to-node cotree branches of order $d \times c$.

The matrix C for an example of network is shown in Fig. 15.1.

A cotree defined as above determines $n + c$ independent loop-sets of network, and each of them is defined by a cotree branch. Each network loop-set includes one cotree branch and a set of tree branches forming a closed loop.

15.2 Current State Vector and Current Flow Method

The current state vector I_m in the above described topological structure includes two subvectors:

$$I_m = \begin{bmatrix} I_n \\ I_c \end{bmatrix} \tag{15.2}$$

where

I_n current vector of node-to-earth cotree branches of order n and
I_c current vector of node-to-node cotree branches of order c.

The current state vector I_m can be derived using general network solution method (see Chap. 11). However, the derivation will need laborious decomposition of b-dimensional equations, which can be substituted by the following simpler derivation.

Let us start from the Kirchhoff's current law for d independent cut-sets:

$$I_d + C_n I_n + C_c I_c = 0 \tag{15.3}$$

where I_d—current vector of tree branches excluding slack node branch of order d.

Using Kirchhoff's voltage law for m loop-sets defined by a node-to-node cotree branches,

$$V_c - C_c^T V_d = 0 \tag{15.4}$$

where

V_c voltage vector of node-to-node cotree branches of order c and
V_d voltage vector of tree branches excluding slack node branch of order d.

Ohm's law for a node-to-node tree branches and node-to-node cotree branches:

$$V_d = Z_d I_d \tag{15.5}$$

$$V_c = Z_c I_c \tag{15.6}$$

where

Z_d matrix of node-to node tree branch impedances of order d and
Z_c matrix of node-to-node cotree branch impedances of order c.

Substituting V_d and V_c from Eqs. (15.5) and (15.6) to the Eq. (15.4), we have the following:

$$Z_c I_c - C_c^T Z_d I_d = 0 \tag{15.7}$$

Substituting I_d from Eq. (15.3) to Eq. (15.7) we have the following:

$$Z_c I_c + C_c^T Z_d C_n I_n + C_c^T Z_d C_c I_c = 0 \tag{15.8}$$

Finally, doing simple algebraic operation, the current vector of node-to-node cotree branches I_c can be found as follows:

$$I_c = -(Z_c + C_c^T Z_d C_c)^{-1} C_c^T Z_d C_n I_n \tag{15.9}$$

In Eq. (15.9), the matrices C_n, C_c and Z_c, Z_d are known from input data. It means that the current vector I_c is a linear function of nodal currents I_n.

Denoting

$$F = -(Z_c + C_c^T Z_d C_c)^{-1} C_c^T Z_d C_n, \tag{15.10}$$

the current state vector I_m is as follows:

$$I_m = \begin{bmatrix} I_n \\ I_c \end{bmatrix} = \begin{bmatrix} 1 \\ F \end{bmatrix} [I_n] \tag{15.11}$$

where vector I_n is a current vector of nodal (node-to-earth) branches, which can be found from nodal loads and nominal voltage of network system.

However, using the physical interpretation of Eq. (15.9) and taking into account the topology of network, the current state vector can be partially found "by inspection" as follows:

Denoting the parts of Eq. (15.9) as follows:

$$Z_F = -(Z_c + C_c^T Z_d C_c) \tag{15.12}$$

$$V_F = C_c^T Z_d C_n I_n, \tag{15.13}$$

the current vector I_c is as follows:

$$I_c = (Z_F)^{-1} V_F I_n \tag{15.14}$$

where V_F is the loop-set voltage vector of order c, and is the analogy to the node-set current vector. Each element of vector V_F is the sum of voltages along the loop-set.

Z_F is the loop-set impedance matrix of order c, which is the number of loop-sets in the classical node-to-node network, and is the analogy to the node admittance matrix. Each diagonal element of a matrix Z_F is a sum of branch impedances forming a loop-set, and each off-diagonal element is a sum of the impedances of the branches common to the ith and jth loop-sets. In real networks, the impedance matrices Z_m and Z_d are non-singular, so the matrix Z_F is non-singular.

It means that the vector Z_F can be found by inspection as follows:

$$I_m = \begin{bmatrix} I_n \\ I_c \end{bmatrix} = \begin{bmatrix} 1 \\ Z_F^{-1} V_F \end{bmatrix} [I_n] \tag{15.15}$$

Finally, the current state vector I_m can be found using either the mathematical (15.11) or physical (15.15) form. Having the current state vector, the current and voltage vectors I and V can be easily found.

The algorithm of current flow method is based on the above Eqs. (15.3)—(15.11). Using the numerical sparse technique, one can substantially reduce the scope of numerical calculations. There is also a possibility of finding the matrix Z_F and the vector V_F by inspection from network graph, subdivided as in Fig. 15.1 into a tree and a cotree.

15.3 Current-Based Load Flow Method

In this section, the current-based load flow method (CBLF) in power system network system is derived, using the above derived current flow method. The starting point is the classical load flow input data:

- Graph of network in the form of a set of b interconnected, closed branches
- Constant voltage value of any one slack node branch
- Active and reactive powers (P, Q loads), and active powers and voltages (P, V loads) of node-to-earth branches except a slack node branch
- Passive parameters of node-to-node branches

The CBLF method uses the iteration process in which the calculation of current state vector is repeated, until initially assumed node-to-earth voltages converge to the right values and reactive power in P, V branches are calculated.

The algorithm of the current-oriented load flow method includes steps before and in iteration process.

1. Choose a tree of a network—including slack node-to-earth branch and any set of node-to-node branches. The remaining branches are a set of complementary cotree.
2. Find incidence matrix $C = \begin{bmatrix} C_n & C_n \end{bmatrix}$ of the tree–cotree structure.
3. Assume a reasonable starting values of reactive power Q in P, V branches.
4. Find current vector I_n of node-to-earth cotree branches using nodal powers and slack node voltage.
5. Find impedance matrices Z_c, Z_d from input data.
6. Find current vector I_c from the Eq. (15.9).
7. Find current state vector I_m from Eq. (15.11).
8. Find current vector of network system I using current flow method.
9. Find voltages of node-to-earth branches.
10. Find the active and reactive power values (P, Q loads) or active powers and voltage values (P, V loads) of node-to-earth branches.
11. Check the voltage mismatches in PV nodes and, if all they are not in the prescribed limits, correct the reactive powers Q in PV nodes.

12. Check the power mismatches and if they are not in the prescribed limits then go to the iteration number 3 else go to the end of iteration process.

The method of obtaining the needed voltage values in PV nodes is based on the evaluation of correction coefficients $\Delta V/\Delta Q$ for each PV node and in each iteration (starting from a second one). According to this coefficient and voltage in P, V nodes, the reactive power Q is changed using accelerating coefficient.

15.4 Verification of Method

In power systems, the load flow calculations are done very often using various methods mainly based on finding voltage state vector. Since publication in 1973 by B. Stott and O. Alsac [ ], the fast decoupled load flow method (based on voltage state vector) and its improved versions are practically used worldwide to this day. The load flow method based on current state vector is a new method and as such needs practical verification on example of real power system network. The current-based load flow method (CBLF) was compared with the classical fast decoupled method (FDLF). Case studies were done on the three test networks as follows:

- Test network I—400/220 kV transmission network, 101 nodes (node-to-earth branches), 6 of them are PV nodes (node-to-earth branches), and 140 node-to-node branches (the real high-voltage power system network)
- Test network II—110 kV network, 77 nodes, 2 of them are PV nodes 92 node-to-node branches—The part of real 110 kV power system network supplied from one 400 kV substation
- Test network III—real 15 kV network, 519 nodes, and 524 node-to-node branches

For each of the test network, the load flow calculations were done using CBLF and FDLF computer programs written in Pascal language. Following results are compared:

- Ite—Number of iterations
- T—Total computation time
- T_i—Time of one iteration
- T_p—Time of preliminary (prior to the iterations) computations

The times (T, T_i, and T_p) in the FDLF method are given in relation to the times in the BOLF method, which are taken equal to 1. The computations were done assuming in both programs the same power accuracy and flat start voltage values.

Case study 1

The load flow calculations for the test network I were done for the following loading states:

- L_{min}—Minimal loading state
- L_n—Normal (average) loading state
- L_{max}—Maximal loading state
- L_f—After failure loading state—the outage of two generating units and four 400 kV lines after short circuit on 400 kV bus bars

The both CBLF and FDLF programs were used, assuming power accuracy $\Delta P < 1$ MW, and in CBLF method, voltage accuracy in PV nodes $\Delta V < 1$ kV. The computation results are shown in the Table 15.1.

The greater differences of a number of iterations and computation times in FDLF method, compared with CBLF method, in loading states L_{max} and L_f, come from the additional iterations needed in FDLF method to correct the Q values in PV nodes (if $Q > Q_{max}$). Such additional iterations are not needed in the CBLF method because to keep the given voltage values in the PV nodes, the corrections of Q values are done in each iteration.

In order to show the convergence ability of both methods, the numbers of iterations were compared for the test network I, by normal loading state L_n, increased by coefficient g to the divergence of iteration process. The load flows in both methods were done assuming all nodes as PQ nodes (in order to avoid the additional iterations due to the necessary changes in Q values in FDLF method). The number of iterations is shown in Table 15.2.

Case study II

The load flow calculations for the test II were done for the following loading states:

- L_n—Normal (average) loading state
- L_{max}—Maximal loading state

Table 15.1 The computation results for CB and FD

	L_{min}		L_n		L_{max}		L_f	
	CB	FD	CB	FD	CB	FD	CB	FD
Ite	8	9	11	13	13	18	14	27
T	1	1.1	1	1.3	1	1.35	1	1.9
T_i	1	0.96	1	0.96	1	0.97	1	0.97
T_p	1	1.34	1	1.35	1	1.70	1	1.75

CB—CBLF method, FD—FDLF method

Table 15.2 The number of iterations

g	1	1.02	1.04	1.05	1.06	1.07	1.076	1.08
Ite FDLF	12	13	15	17	20	31	DIV	
Ite CBLF	10	11	12	13	14	22	33	DIV

DIV—Iteration process diverges

Table 15.3 The results of computations for CB and FD

	L_n		L_{max}		L_f	
	CB	FD	CB	FD	CB	FD
Ite	3	9	5	12	10	15
T	1	2.52	1	2.97	1	1.98
T_i	1	1.18	1	1.20	1	1.21
T_p	1	2.14	1	1.86	1	1.51

CB—CBLF method FD—FDLF method

Table 15.4 Characteristic feature of the CBLF and NDLF method

	Normal load		High load	
	CBLF	NDLF	CBLF	NDLF
T	1	3.53	1	5.7
T_i	1	1.22	1	1.24
Ite	2	5	3	7

- L_f—After failure loading state—normal load state after the outage of one generating station (40 MW) and four 110 kV lines due to the short circuit on 110 kV bus bars

The results of computations, shown in the Table 15.3, enable the comparison of both methods.

The greater numbers of iterations and computation times in FDLF method come mainly from the difference of an order of matrices to be inverted. In the CBLF method, the order of a mesh impedance matrix Z_F (see Eq. (15.12) is equal to $c = 14$, while in FDLF method, the order of Jacobian matrix is equal to $n = 76$.

Case study III

The load flow calculations for 15 kV test network are done using COLF method and the Newton's decoupled load flow method (NDLF) because FDLF method diverges. The results are given in Table 15.4.

Table 15.4 shows the characteristic feature of the CBLF method. Its effectiveness rises with a lower ratio of the number of node-to node branches to the number of nodes, which in this case is 5/519 = 0.01. In CBLF method, the total time of computations and the number of iterations are above two times than in the NDLF method. So, the CBLF method is more effective.

15.5 Conclusions

From the theoretical viewpoint, the CBLF method is based on a non-conventional approach to the load flow solution. As such, it differs substantially from the other methods. Compared with recently used methods, the main differences of the CBLF method are as follows:

- From the mathematical viewpoint, CBLF is a linear method because the main computation problem is finding the current state vector from a set of linear equations, while in recently used methods, the main problem is finding a voltage state vector from a set of nonlinear equations.
- Generally, CBLF method is numerically more stable, compared with classical methods. Finding load flows in more loaded networks needs less iterations.
- The topological structure (tree and cotree) of a network graph in CBLF method is different than in classical methods. The tree of network, instead of node-to-earth branches, consists of some node-to-node branches and slack node-to-earth branch. Remaining branches including node-to-earth branches are cotree branches.
- The final results of computations, apart of voltages and load flaws, include additionally current flaws in a network.

  From the practical viewpoint, the main differences of CBLF method are as follows:
- Instead of solving, in each iteration, a set of n (number of nodes) equations, the set of c equations is solved. It makes the effectiveness of COLF method depending very much on a coefficient k

$$k = c/n \tag{15.16}$$

where c is the number of node-to-node branches in a cotree and n is the number of node-to-earth branches mines one in the network.

- In power system networks, the effectiveness of CBLF method depends on the voltage level. For a high-voltage networks (400 and 220 kV) in normal load conditions, the effectiveness of CBLF is nearly the same as the classical methods; however, in abnormal conditions, it may be much higher (see Tables 15.1 and 15.2). In cases of voltage levels less than 400 kV, CBLF method is faster then the classical methods (see Tables 15.3 and 15.4).
- As to the disadvantages, the PV nodes in the CBLF method must be specially treated. In the iteration process, they are substituted by PQ nodes, and in order to keep the voltages on the prescribed values, the corrections of reactive power Q in PV nodes are needed in each iteration. The algorithm of correction (used in the case studies 1 and 2 shown above) is based on the evaluation of V/Q sensitivities and the application of an accelerating coefficient. The correction algorithm used is optimal (minimum number of iterations) for the case studies above but it may be not optimal for other network structures.
- The CBLF method is universal. It can be effectively used for all kinds (voltage levels) of a meshed and open power system networks.
- From the computational viewpoint, the CBLF computer program needs less computer space than the recently used load flow programs.

Zeitfracht Medien GmbH
Ferdinand-Jühlke-Straße 7
99095 Erfurt, Deutschland
produktsicherheit@kolibri360.de